PLATE 4

The famous baobab outside the Commissioner's office at Katimo Malilo in the Caprivi Strip fitted out with a flush toilet. Photograph by D. Killick, Dec. 1958.

The Baobab—Africa's upside-down tree*

G. E. WICKENS

Summary. This is an attempt to pull together what is known about that extraordinary tree, the African baobab (*Adansonia digitata* L.—*Bombacaceae*). There are many surprising gaps in our knowledge, which are most likely to be reduced by closer collaboration between fieldworker, laboratory and herbarium botanist.

CONTENTS

Accepted for publication October 1981

*A few extra copies of this article have been printed and can be obtained separately from the Royal Botanic Gardens, Kew, Richmond, Surrey, England; price £3·90 or £4·06½ by post inland or surface post overseas

PART I

1. Introduction

There can be few people who have travelled through the savannas of tropical and southern Africa and not seen and recognized the baobab, *Adansonia digitata*. With its enormous, swollen trunk and grotesque outline it is easily recognized. Its sheer size and massiveness suggests great age and historical significance. It is a tree that we should know well—yet how well do we really know it?

A great deal of information, often of a rather regional nature, is contained in the earlier reviews by Gerber (1895), Chevalier (1906) and Adam (1962). This review results from a further survey of the literature and is an extension of the Baobab Map Project initiated by Lucas (1971). That project showed how different a distribution map drawn from the personal observations of many collaborators was from one based on museum records alone. It now seems useful to present a summary of all we know about the baobab throughout its range and to highlight any gaps in our knowledge—and for such a well-known tree there are surprisingly large gaps—in the hope that further fieldwork may be stimulated.

If our knowledge is still so imperfect regarding such a familiar tree as the baobab then it is rather frightening to consider how even less is known about other constituents of the African flora. The consequences of well-intended but misguided land development schemes based on inadequate knowledge of plants, either as individual species or as communities, could in the long term, prove disastrous.

2. Historical background

(a) *Fruits and the name*

Baobab fruits were apparently known to the ancient Egyptians, although the tree is not native in Egypt. The fruits have been reported from their tombs but unfortunately no-one recorded in which the fruits were found, and the Museums of Paris and Turin, said to have been the depositories for this material, cannot confirm that the fruits still exist. Such fruits must have been rare; there is no mention of the baobab in any of the more recent works on plants in ancient Egyptian tombs.

It is said that the baobab (? fruit) was referred to in the inscriptions near Aswan of a caravan-leader named Harkhuf (c. 2500 BC), although present-day translations make no mention of it. Nevertheless, it is likely that caravan leaders could have brought the fruit from the Sudan, where the tree occurs as far north as Abu Haraz on the Blue Nile (south of Khartoum), and eastwards through Kasala to Massawa on the Red Sea (Map 1). This boundary certainly lies within the area penetrated by the caravans from Egypt. We can speculate that the fruit may have been introduced for its medicinal properties as a febrifuge, which is the reason for its later presence in the Cairo markets during the sixteenth century. The fruit is apparently not sold in the herb and spice markets of Cairo today.

In 1592, the Venetian herbalist and physician Prospero Alpino wrote that the fruit was known in those markets under the name *bu hobab*, which gave rise to the common European name *baobab*. One Cairo botanist, the late

Mohammed Drar, suggested a derivation of *baobab* from the Arabic *lobab* or *lobb* referring to the fruit pulp, once used in medicine. However, the Arabs do not nowadays call the tree by this name, and I believe a more likely suggestion is *bu hibab*—'the fruit with many seeds'. Trees cultivated in Egypt are known as *habhab*. In the Sudan the common Arabic vernacular name is *tebeldi*, more rarely *homeira* or *humr* on account of the reddish tinge to the bark; the fruit is known as *gongoleis*. Thus, it would appear that *bu hobab* was a local name invented by the Cairo merchants for a fruit (and tree) which they did not know in the wild.

The fifteenth-century Portuguese travellers to West Africa knew the fruit as *cabaçevre*, and the seventeenth-century French as *calebassier*; it was the French botanist Michel Adanson who reapplied Alpino's name of *baobab* to the fruit and later to the genus itself. He modestly rejected the Linnaean name of *Adansonia*, coined in his honour by his teacher, Bernard de Jussieu, after publication in 1757 of the results of his exploration of Senegal; however, in obedience to the rules of nomenclature of plants, *Adansonia* is nowadays the accepted scientific name. The statement by Linnaeus that the habitat was 'Egypt' is obviously a reference to Alpino, nevertheless it is clear that he also knew of its presence in Senegal through Jussieu.

Curiously, in Mauritania the Arabic name is *teidoûm*, which is of Berber origin, although the berbers of North Africa are unlikely to have known the tree.

In Europe too *Adansonia digitata* was first known by its fruit. The earliest recognizable description appears to be Scaliger's in 1557 under the name of *guanabus*, although the credit is usually given to Alpino, who provided an accurate illustration of the fruit and a rather imaginative picture of the flower and leaves. In 1605 Clusius published excellent illustrations of both fruit and leaves. Other early (pre-Linnaean) authors included Jean Bauhin, Cherler & Chabrey (1650–51), Caspar Bauhin (1671) and John Ray (1688).

(b) *Early field observations*

For such a conspicuous and distinctive tree the early literature references to field observations are surprisingly sparse. The first recognizable reference is that by the great Arab traveller Ibn Batuta, who, in 1352, commented on a weaver in Mali weaving cloth within the shelter of a hollow trunk. Other hollow trunks were being used for water storage; *baobab* water-tanks are well-known to this day, especially in the Sudan Republic.

The *baobab* was also known to the Moroccan scholar Leo [John] Africanus (1600), who between 1511 and 1517 travelled widely through northern Africa.

Recognizable descriptions were given by Portuguese navigators such as Gomes Eannes de Azurara for trees seen on Bisiguiche (Bisiguienne), off Guinea in 1447–48, by Bento Banha Cordoso in 1622, as well as from the mouth of the Senegal by the Venetian Aloysius de Cada-Mosto in 1454. It was also from Senegal that the use of hollow trees for burial purposes was first depicted by Dapper in 1686.

3. Baobab relatives

The African baobab and its allied species are members of the small pantropical family *Bombacaceae*, which also contains the economically important kapok tree *Ceiba pentandra* (L.) Gaertn., balsa wood tree *Ochroma pyramidale* (Cav. ex Lam.) Urb. (syn. *O. lagopus* Sw.) as well as the durian, *Durio zibethinus* Murr. with its evil-smelling yet exquisitely flavoured fruit. Although a number of genera have soft, water-storing woody tissue, none quite attain the monstrous proportions of *Adansonia*.

The generic limits of the family are somewhat uncertain, with some genera exhibiting quite strong affinities with the *Malvaceae* or *Sterculiaceae*. Some 29 genera and c. 225 species are currently recognized. These are divided into 6 tribes, 4 of which are restricted to tropical America, 1 to Australasia (7 genera and c. 50 species) and the remaining tribe, the *Adansonieae*, is pantropical.

Four genera of the *Adansonieae* are restricted to tropical America; *Bombacopsis* is mainly in tropical America but with 1 or more species in tropical Africa, and Bombax extends from India to New Guinea. Finally, *Adansonia* with about 9 rather similar species occurs in tropical Africa *(A. digitata)*, western Madagascar and western Australia *(A. gregorii)*. This pattern of species distribution is in itself most unusual; plants of western Australia have a great deal in common with those of the Cape Province of South Africa, but not often with plants from tropical Africa. The distribution history with its geological implications are discussed in more detail on p. 200.

4. Description of the Baobab

The following description is based largely on those of Wild (1961), Robyns (1963), Breitenbach (1965), Villiers (1975) and Palgrave (1977) (see also Fig. 1).

Massive deciduous tree, usually not more than 20 m tall but can reach 23 m. Trunk stout, tapering or cylindrical and abruptly bottle-shaped or short and squat, up to 10 m in diameter (Plate 5). Bark smooth, reddish-brown, greyish-brown or purplish-grey. Crown large, usually spreading; primary branches either well distributed along the trunk or confined to the apex, stout but gradually tapering, young branches often tomentose, rarely glabrous.

Leaves simple or digitate, alternate at the ends of the branches or borne on short spurs on the trunk (it is uncertain whether the latter condition is normal or whether a result of stripping the bark). Leaves of young trees usually simple. Adult trees begin each season by producing simple leaves followed by progressively 2–3-foliolate leaves which are apparently early deciduous; mature leaves 5–7(–9)-foliolate and c. 20 cm in diameter. Petiole up to 16 cm long, densely pubescent becoming glabrescent or rarely glabrous. Leaflets sessile or shortly petiolulate, elliptic, oblong-elliptic or obovate-elliptic, 5–15 × 1·5–7 cm, apex acuminate, mucronate, base cuneate, decurrent, margins entire, stellate-pubescent beneath when young but soon glabrescent or rarely entirely glabrous; stipules early caducous, subulate or narrowly triangular, 2–5 mm long, glabrous except for the ciliate margins.

Flowers pendulous, solitary or paired in the leaf axils, large, showy. Pedicels tomentose, up to 20 cm long in southern and eastern Africa, up to 90 cm long in West Africa and Angola; bracteoles 2, small, early caducous, borne near the apex of the pedicels; bud ovoid to globose, apex conical to apiculate. Calyx 3–5-lobed, 5–9 × 3–7 cm, divided to ½ or ¾, shortly

FIG. 1. *Adansonia digitata*. **A** habit × ⅔; **B** dissected bud × ⅔; **C** flower × ⅔; **D** anther × 4; **E** fruit × ⅔; **F** seed × 2. **A** from *Newton* 727, **B–F** from *Freeston* 35.

tomentose outside, velvety pubescent inside, lobes oblong to broadly triangular, 3–7 × 1·8–2·5 cm, apex obtuse to subacute. Petals 5, white, overlapping clockwise or anti-clockwise, even on the same tree (Davis & Sukhendu 1976), very broadly obovate to oblate, 5–10 × 4·5–12 cm, apex rounded, base shortly clawed, very sparsely hairy but densely hairy on the claw inside or glabrous. Stamens very numerous (720–1600 fide Davis & Sukhendu 1976), united below into a staminal tube 1·5–4·5 cm long, free portion of filaments slender, equalling the staminal tube and reflexed to form a ±horizontal ring; anthers ±reniform, 2 mm long. Ovary 5–10-locular becoming unilocular above, conical to ovoid-globose, 1–2 × 1–1·2 cm, silky tomentose, placentation parietal, with repeatedly dichotomous funicules (Heel 1974); style exserted c. 1·5 cm beyond the anthers, reflexed or erect, hairy near the base; stigma with 5–10 fimbriate-papillose lobes up to 8 mm long.

Fruit variable (Plate 6), globose or ovoid to oblong-cylindrical (in Angola), sometimes somewhat irregular in shape, 7·5–54 × 7·5–20 cm (up to 25 cm long in southern Africa, 40 cm in West Africa and 54 cm in Angola), apex pointed or obtuse, rarely sulcate, covered by a velvety tomentum of pale, yellow-brown hairs; ripe fruits ±filled with a dry, mealy pulp; seeds many, embedded in the pulp, ±reniform, 1–1·3 × 0·8–1 bm, testa smooth, dark brown to blackish.

Seedlings with flattened hypocotyl, 4·2–6·4 × 3·3–5 cm, entire; epicotyl 3–4·8 cm long; tricotyly, tetracotyly, hemitricotyly and hemitetracotyly in seedlings have been reported by Srivastava (1959). First leaves petiolate, generally simple, narrowly linear.

This description of the more obvious features covers the whole of the species throughout its native range in Africa. Within that range there is a great deal of variation, so much so that, when the trees are better known, we may find ways of distinguishing consistent variants which could then be named for convenience of reference. The subject is discussed in more detail on p. 198, where details of cytology and chromosomes, pollen and anatomy are also given.

5. Natural distribution and ecology

The baobab occurs in most of the countries of Africa (Map 1) south of the Sahara except for Liberia, Rio Muni, Burundi, Uganda and Djibouti; it barely enters eastern Chad and in South Africa is only known from the Transvaal. It has been introduced into Gabon, Central African Republic, the central Zaïre basin and possibly elsewhere. It has been recorded from the islands of Sao Tomé, Principe and Annobon, but whether they are native or introduced by man is uncertain. Even in countries where it is known to be native there are known examples where the baobab has been planted by man as an amenity tree, either within its natural range of distribution as at Messina in the Transvaal or outside its range as at Khartoum in the Sudan Republic.

We know all too little about the habitat requirements of the baobab, although there is a good regional account for southern Africa by Werger (1978) & co-authors. Basically the baobab is a characteristic representative of the drier plant communities of the Sudano–Zambesian lowlands with c. 200–800 mm annual rainfall, with extensions into the higher rainfall areas receiving up to 1400 mm rainfall. It is possible that in these higher rainfall areas its natural distribution has been extended by man.

However, these generalizations do not tell the whole story, for there is evidently a good deal of variation caused by interaction of environmental factors. The ecology of the baobab has not been fully worked out. What details we have are summarized here under regional headings.

(a) *Western Africa*

The baobab is widely distributed through the savanna region of West Africa, especially in the drier parts, with extensions into the forests, where it is usually associated with habitation. In a densely populated area such as West Africa it is virtually impossible to know whether such intrusions into the forest are due to man or whether they represent natural enclaves of savanna vegetation within the forest boundary that have attracted occupation by man.

Baobabs are present in the Sudan zone streamside vegetation of Nigeria, presumably a natural habitat. It is also a characteristic relic tree of the permanent Sudan zone farmlands, where it occurs in association with *Parkia clappertoniana*, *Butyrospermum paradoxum*, *Acacia albida*, *Tamarindus indica* and *Balanites aegyptiaca*. Characteristically it occurs on free-draining sandy textured soils, but can tolerate shallow lateritic soils as well as being found on rocky hillsides, moist lower slopes and around the margins of seasonal pools, but rarely on poorly drained sites.

It has been suggested by a number of authors, that the close association of the baobab (and tamarind) with villages indicate that the tree must have been introduced by man. This may certainly be true for some sites but I am equally convinced that there are also many villages that have been sited so as to take advantage of the existing shade of an indigenous baobab.

Further to the north of Nigeria, near Lake Chad, the baobab is a characteristic tree of the Quarternary dune system overlying the clays of the former Mega–Chad basin. It does not occur on the seasonally inundated interdunal clays. Associated trees and shrubs include *Acacia senegal, A. albida, A. seyal, Bauhinia rufescens* and *Indigofera arrecta*. Again the baobab is often associated with village sites; I suggest that the villagers took advantage of the shade and economic potential of existing trees when siting their villages as well as avoiding the surrounding clays. The baobab here occupies precisely the same ecological niche above the clays as it does further to the east in Kordofan Province of the Sudan Republic and usually without an associated village site. There is also some historical evidence to support this idea in that Kuka, the former Kanuri capital in northern Nigeria, was so named because the city was founded in the shade of a *kuka* tree, the Kanuri and Hausa name for the baobab.

Whereas large trees are reported from the Nigeria–Niger border, stunted trees were noted in the Zinder area (Niger) where it apparently reaches its northernmost limit in an area receiving c. 550 mm annual rainfall. A more reasonable minimum rainfall requirement for West Africa generally would, from the distribution map (Map 1) appear to be in the region of 300 mm with an absolute minimum of c. 90 mm in Mauritania.

In the former French territories of West Africa, the baobab seems to be definitely 'calciphile' on account of the large and dense populations occurring on the calcareous soils of Cape Verde and in the region between Kidira

(Senegal) and Kayes (Mali), with isolated or more extensive occurrences on the estuarine and lagoon shell deposits of the Senegal River. It is only sporadic in occurrence along the edges of lateritic carapaces, along tracks or in the open savanna.

The baobab has also been recorded from the coastal plains of Ghana, Togo and Dahomey in areas receiving between 700 and 800 mm rainfall per annum. It also occurs south of the equator on the coastal plains of the Congo, Kubango and Kunene, which has led to speculations on a possible maritime introduction by man. Baobabs are also found in the secondary *Heteropogon–Hyparrhenia* grasslands of the coastal regions of the Bas–Zaïre, and it occurs in clumps, along with the obviously introduced mangoes *(Mangifera indica)* and cashew trees *(Anacardium occidentale)* on sites of former habitation. Further field and cytological work is needed to discover the origins of these coastal populations.

(b) *Northeast Africa*

The baobab has not been recorded from Chad east of the Chari but re-occurs in western Sudan. I believe this to be a true absence and not due to inadequate observations.

The most westerly record in the Sudan is of a small streamside grove to the north of Zalingei in *Anogeissus* savanna woodland. A few trees occur in the eastern foothills of the Jebel Marra massif at altitudes between 900 and 1250 m but it is on the basement complex soils of the plains at 500–600 m, to the east of the volcanic soils of Jebel Marra, where the baobab becomes increasingly more frequent, suggesting that the Nile/Niger divide is in fact a natural barrier for the eastern population of the species.

As in West Africa, the baobab shows a preference for free-draining soils, preferably with a sandy topsoil overlying loamy substrates. It does not occur in areas of deep sands, presumably because it is unable to obtain sufficient anchorage and moisture. Similarly, in a dune system it occurs on the lower slopes bordering the interdunal clay deposits. Occasional baobabs are also to be found on the crests of quartz ridges, presumably where the seed has been brought by baboons, etc.

The northern limit is in the semi-desert scrub and grassland communities to the 100 mm isohyet, with stunted trees fringing the seasonal drainage lines; better grown and often large groves of baobabs occur in the fossil drainage systems where there is probably a fairly high water table.

The baobab becomes increasingly more abundant south of the 200 mm isohyet, in the *Acacia* thorn savannah and scrub communities, on sandy and basement complex soils, and extending into the deciduous savanna woodlands. On the hill slopes of the Nuba Mountains the baobab is often associated with seepage lines. Throughout its distribution in the Sudan I could find no regeneration except in the Nuba Mountains (rainfall 700–800 mm). The regeneration might be due to the higher rainfall although it is suspected that protection from fire might be an important factor too.

In Ethiopia the baobab is restricted to the lowlands of Eritrea. It has been fairly extensively reported by Italian botanists from the coastal plains of southern Somalia, although it is now disappearing from many of these localities.

(c) *East Africa*

There are no reports from Uganda, and in Kenya it is found below the

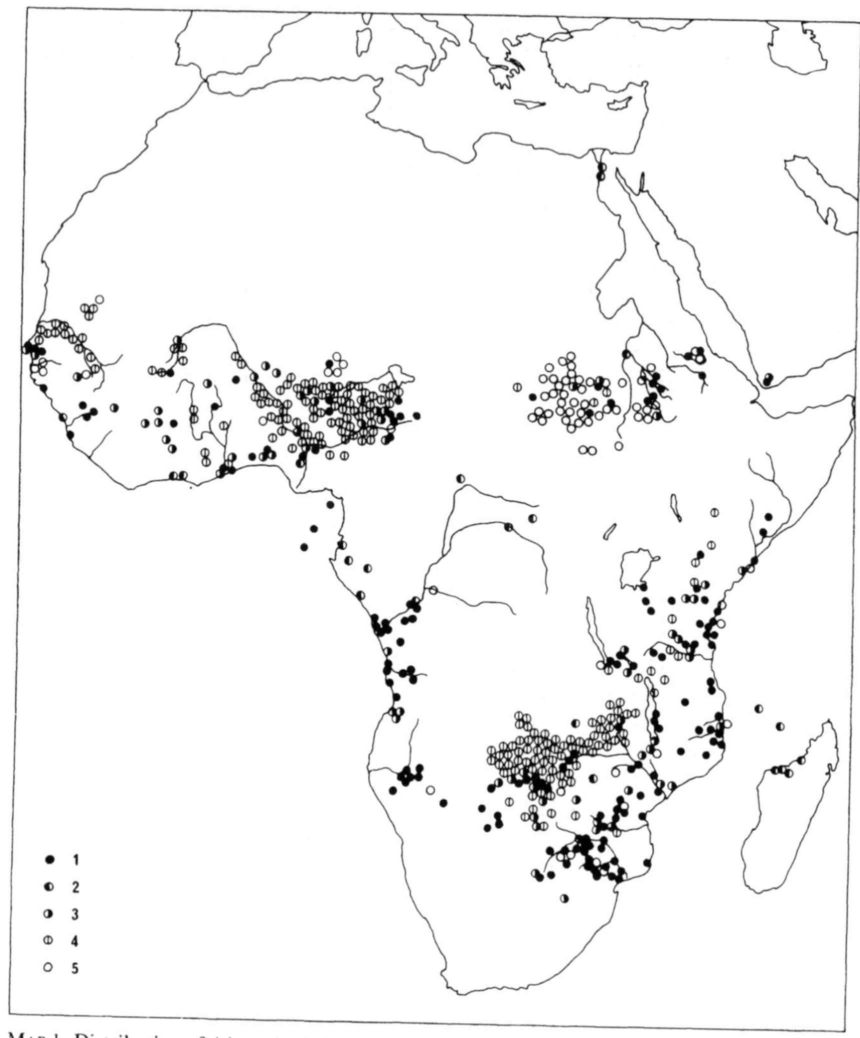

MAP 1. Distribution of *Adansonia digitata* in Africa and neighbouring areas. **1** Distribution based on Herbarium and flora records; **2** Specimens known to be cultivated or introduced; **3** Distribution based on published and unpublished photographs; **4** Distribution based on the Kew 'Baobab Survey' information; **5** Records obtained from travel literature, maps etc.

1000 m contour, chiefly in *Acacia–Commiphora* bushland and scrub and the semi-desert grasslands of northern Kenya in areas receiving c. 250 mm rainfall. In Kenya, the baobab has been recorded from the coastal coral, on breccia in association with *Ehretia petiolaris, Fagara chalybea, Grewia plagrophylla* and *Heeria mucronata*, bushland, while in Tanzania and Mozambique, it is a conspicuous emergent of the *Afzelia guineensis* coastal forest, especially on outcrops of coral limestone. The rainfall is in the region of 700–1400 mm per annum. Elsewhere in Tanzania it is a conspicuous emergent in many parts of the deciduous bushlands or a conspicuous relic in areas of upland plateau cleared for cultivation; scattered baobabs are also found along the valley edges of the central plateau east of the Congo–Wembere divide.

(d) *Southern Africa*

In the valleys of the Zambezi and its major tributaries, such as the Luangwa, etc., baobabs occur as a constituent of the mopane *(Colophospermum mopane)* woodland on relatively heavy-textured soils. The mopane woodlands occur at altitudes ranging from 200–500 mm in the Tete District of Mozambique to 1000 m on the Kariba basin and have an annual rainfall of 500–700 mm.

In the somewhat poorly drained soils of Zimbabwe the baobab is found in the savanna with *Cordyla africana, Kigelia africana,* etc., also in the dry woodlands of Malaŵi receiving c. 700 mm rainfall as well as in the poorly drained plains of the Zambezi delta. It is also reported from the mountainous parts of the Kaokoland of Angola and South West Africa, where it occurs in the mopane woodland. In the northern and central Angolan plateau the baobab occurs in the thicket and savanna vegetation in areas receiving upwards of 1300 mm rainfall as well as on the drier coastal lowlands.

In South Africa the baobab is a conspicuous constituent of the Limpopo basin, especially in the hot, dry, frost-free sandy country to the north of the Zoutpansberg mountains and the Olifants River in the east. A few stragglers grow further south, including some big trees on the southern banks of the Nwaswitsontswe River, some 80 km to the south of the Olifants River, a few in the Waterberg and one in the Rustenberg district.

The baobab is at its best on deep well-drained soil at altitudes of between 450 and 600 m above sea level with a rainfall between 300 and 500 mm per annum. This would seem to apply for its distribution throughout tropical Africa, similarly in southern Africa except that here frost becomes the limiting factor.

6. Occurrence outside Africa

In Arabia the baobab has been recorded from Samsara in North Yemen and in Oman, from the Wadi Hinna, near Mirbat in Dhofar and west of the Wadi Hinna at Dhalqut. Popov & Zeller (1963) seem to suggest that the presence of *Adansonia* and other African species in the south-west corner of Arabia is natural; they also imply that there are a number of baobab trees present. However, only the one specimen is known from Samsara, although it is possible that the trees have been destroyed or that they may exist in the now inaccessible parts of the South Yemen. It is not certain whether the Dhofar specimens are native or introduced by Arab traders, the large community at Dhalqut certainly looks like a long-established and naturally regenerating population.

Settlers from Zanzibar are thought to have taken *Adansonia digitata* to Madagascar.

The baobab is fairly widely distributed in India and its distribution coincides remarkably well with the areas under Moslem control and it must therefore have been introduced by man early in the thirteenth century at the start of the African slave trade. It cannot be identified in any of the ancient Sanskrit writings and it seems to have no Sanskrit name. One suggestion is that the baobab can be identified with the mythical wishing-tree *Kalpa-vriksha,* but illustrations of this are so stylised that one cannot be sure, and others maintain that *Kalpa-vriksha* is more likely to be the banyan, *Ficus benghalensis.* Perhaps its bizarre appearance was sufficient reason for its introduction; the tree is not generally utilized (except ocassionally for rope making) yet, since the baobab in India is so often associated with temples and shrines, and is the subject of religious ceremonies, it could be argued that the temples followed the trees, as in Kuka in northern Nigeria, rather than the reverse. If so, such trees would represent either an early introduction during the Krishna period, c. 2500 BC, or even be native and link up with the Arabian specimens. Vaid (1978a) does accept that some trees were introduced by Arab, French or Portuguese traders. Certainly Arab traders introduced the baobab into Ceylon.

The baobab has also been introduced into many parts of the tropics and subtropics as an ornamental, including the gardens of Cairo (where they are believed no longer extant), Mauritius, where it may once have been naturalized, Malaya, Java, New Caledonia, Hawaii, Philippines, West Indies, the Antilles and Guyana. French immigrants are said to have introduced it to Cuba from Haiti and in 1912 seed from Cuba was sent to the U.S.A., from which trees have since become established in Florida.

It is odd that all reports of ornamental baobabs grown around the tropics refer to the African species, not one of the Madagascan ones. Some of these have striking bottle-shaped trunks eminently suitable as curious ornamentals. One wonders why—or even if, perhaps, these ornamentals have not always been correctly identified.

7. Natural history

(a) *Growth and development*

Seedlings and saplings of baobabs in the wild are often not recognized, since usually they lack the characteristic swollen trunk and palmately digitate leaves of the adults. Saplings often retain their simple seedling leaf form for many years. However, observations on pot-grown plants have shown that the variation in the shape of the leaf can occur from a very early age.

The first simple leaves of the seedling can change drastically with sixth leaf being digitate or gradually with irregular teeth after the first ten to twelve simple and entire leaves, followed by progressive foliation after the twelfth node. The simple leaf may even persist for years, occurring in combinations of simple, 2-foliolate and 3-foliolate leaves and variations of these up to the 5–7 leaflets of the adult tree. My own observations in the Sudan are that even mature trees follow this sequence each year; the annual flush begins with simple leaves although these early, 'primitive' leaves soon fall and are replaced by the familiar digitate ones. Unfortunately, no observations were made on leaf-fall etc. except to note an apparent lack of 'seasonality'. In Kenya, however, the leafy period appears to be confined to the rainy season. The consequent limitation on photosynthesis in the leaves is probably offset somewhat by the possession of a distinct green layer just below the waxy outer surface of the bark.

As the trunk grows and thickens with increasing moisture content three stages of growth can be recognized. At first the stem has a steady, steep taper with a little branching until more or less adult height is reached (Plate 5B). A rapidly growing young tree can be a great deal wider at the base than at 0·3 m, above which it will produce a steady taper; a slow growing tree lacks the basal thickening. During the second stage the main trunk is of more or less uniform diameter throughout except for a constriction below the branches, becoming 'bottle-shaped'. Finally a widespread branching crown is produced, the trunk increases in diameter and becomes hollow and the root system extensive.

This is possibly an over-simplification and may only apply to observations in East Africa where the 'bottle-shaped' baobab appears to be a fairly common form apparently not reported elsewhere. Further field observations are clearly required. There is another form with a short and remarkably thick-set and swollen trunk which is subdivided not far above the ground into a number of massive branches; there are also forms with an unbranched, cylindrical trunk and with a trunk bearing branches for at least two-thirds of its height. These latter forms may perhaps be regarded as variations or intermediate stages. There is also the observation that young saplings may sprout into 2 or 3 slender trunks which slowly increase in girth and after some fifteen years coalesce to form a 'gouty' tree.

In young trees the branches tend to be more erect; they spread and droop with age; this 'middle-age' spread seems to develop between 30 and 40 years of age; thereafter the trunk thickens rapidly, attaining 4·5 m in diameter after 100 years. The strength of the branches is almost entirely due to their water content, in fact so much so that during periods of severe drought the branches may become so dehydrated that they snap with a loud report, occasionally bursting into flames.

Fallen trees often survive and may even produce erect, 'baobab tree-like' growth from the prostrate parent. Even more bizarre is the ability of felled trees to grow a bark that covers the exposed stump and produce a new shoot from the centre of the stump in addition to peripheral shoots (Plate 7A). Perhaps a regenerative bud is carried by the bark to the centre instead of arising from the centre—an unusual possibility.

The seedlings have a swollen, turnip-like taproot, a feature that is retained in young trees. How soon trees develop their extensive system of massive subsurface lateral roots is not known, but at some stage in their development the taproot must wither away for I have examined numerous uprooted trees and found no suggestion of a taproot.

(b) *Age determination*

The enormous girth of the mature tree has led observers to assume that the baobab lives to a great age. There have been many attempts to assess the age of the tree on the basis of incremental growth from girth measurements or by the counting of annual rings (see below). Girth measuring is not as simple as it appears, since the bole is often so irregular that the girth is difficult to measure accurately. Variation in water content, furthermore, causes fluctuation in girth; the trunk of a young tree of 1 m in diameter can expand so rapidly after good rains that the soil is displaced and heaped up around the base of the trunk. Conversely, transpiration can be so rapid and the trunk shrink so much that a gap of over 1 cm can appear in the soil surrounding the trunk, sometimes sufficient to insert the hand. Even long-term girth measurements have sometimes registered such shrinkage; recently it has been suggested that volume increment is a more reliable guide. Counts of the annual rings from trees of known age have been within 2% of the known age. The counting of annual rings using a core extractor is feasible for trees up to 1·5 m in diameter, but there are severe technical difficulties with larger diameters.

The first attempt to calculate the age of the baobab is attributed to Adanson. In 1749 he examined two trees on the Magdalene Islands off Cape Verde; the trees, which have now disappeared, were inscribed with the names of European sailors who had landed there during the fifteenth and sixteenth centuries. Adanson calculated that the trees were 5150 years old. This aroused the wrath of David Livingstone, a follower of Bishops Usher and Lightfoot, who had calculated that the world had been created in 4004 BC; Adanson's calculations implied that the baobab must have been alive before the Flood, ergo, no Flood! However, Livingstone, using growth rings, did arrive at a figure of over 4000 years for his oldest tree!

More recently, G. L. Guy, a retired forester, has painstakingly traced a number of trees whose girth measurements have been recorded by early travellers in southern Africa, as well as the more recent records by foresters, including a well documented sample plot near Messina in the Transvaal. He concluded that the growth rate is slow and that some trees, such as the famous Livingstone Tree at Shiramba on the lower Zambesi River, must be over 2000 years old.

The often-cited carbon dating of a tree 4·5 m in diameter from the Kariba Dam site on the Zambesi River gave an age from the heartwood of 1010 ± 100. The average annual increment in radius over the last 1·14 m of the total radius was 1·5 mm, which agreed reasonably well with the measured average ring width over the last 20 cm of growth of c. 1·1 mm. With diameters in excess of 10 m, ages of c. 2000 years do not appear excessive, although the largest trees are not necessarily the oldest.

(c) *Flowering*

There are mixed reports about flowering, as there are for the vegetative growth. From West Africa there is a rather surprising report that baobabs flower and fruit at eight to ten years. South African reports suggest that trees cultivated at Messina started to flower when 16–17 years old while in Zimbabwe first flowering has been suggested at 22–23 years. This may be confirmation that the geographically isolated West African baobabs are really rather different from those of East and southern Africa—or it may be a reflection of differences in climatic regimes. Flowering is said to occur just before or at the start of the rainy season, although my own observations in the Sudan suggest that flowering may occur throughout the year except at the height of the dry season—January to March.

The buds begin to open from the late afternoon to soon after sunset and are fully open by the following morning; the calyx and corolla lobes curl back to expose the stamens (Plate 8B). During the next morning the calyx and corolla begin to straighten and re-envelop the stamens. This is accompanied by a progressive wilting of the flowers until the late afternoon when the withered corolla becomes detached and slowly slides down the staminal tube and hangs limply over the stamens (Fig. 2). Thus, the lifespan of the flower is not more than 24 hours with a probable pollination period of 16–20 hours.

(d) *Pollination*

Because of their structure, the Viennese biologist Otto Porsch, who was at the time (1935) working in the Botanic Gardens, Buitenzorg (now Bogor, Java), suggested that the pendulous flowers of baobabs were adapted for pollination by bats. In the following year this was confirmed by Pijl who

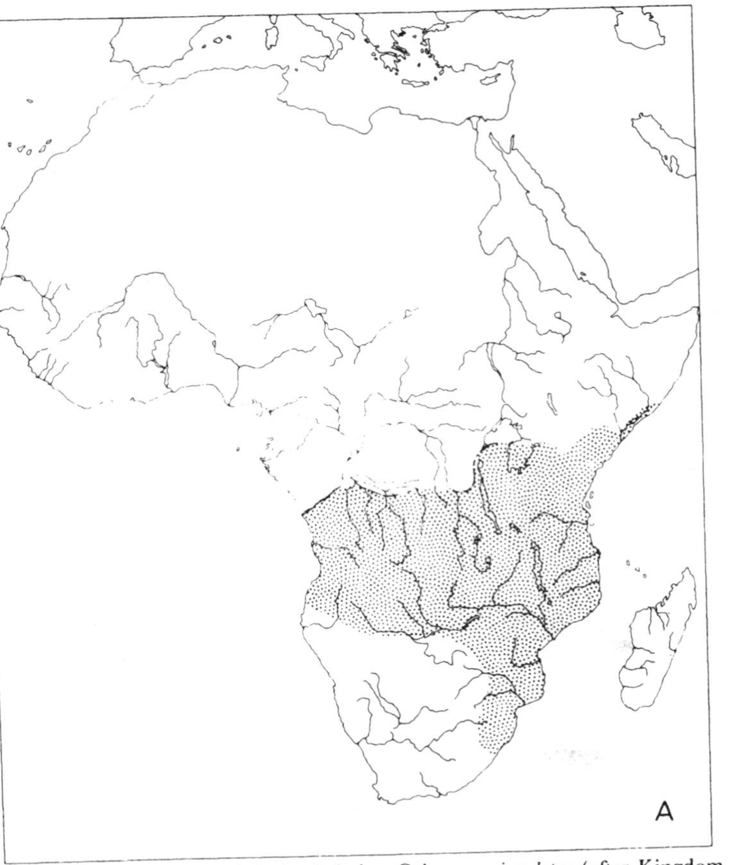

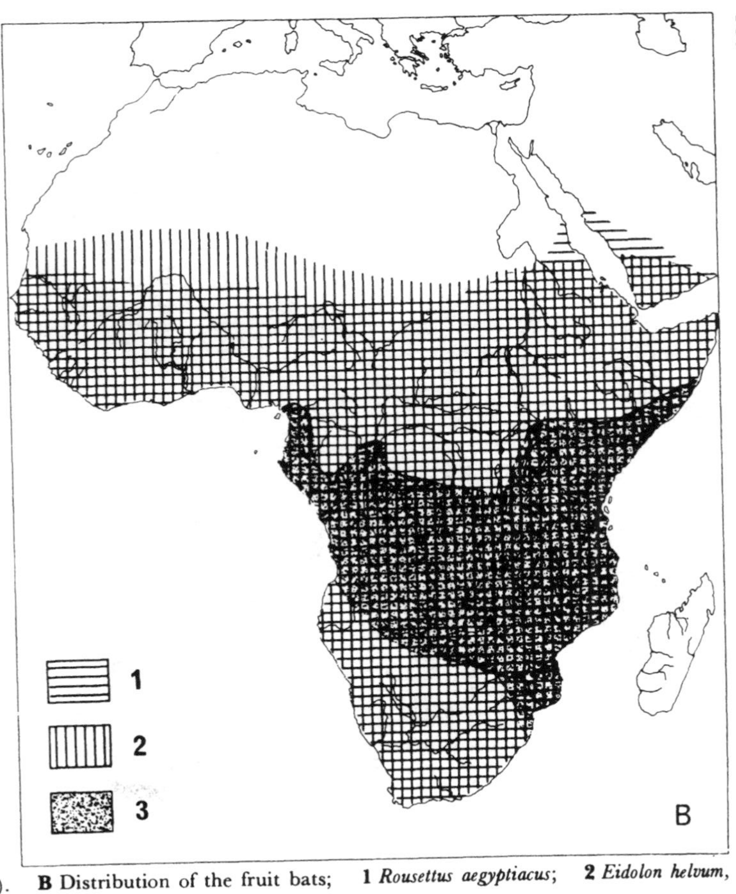

MAP 2. **A** Distribution of the bush baby, *Galago crassicaudatus* (after Kingdom, 1971). **B** Distribution of the fruit bats; **1** *Rousettus aegyptiacus*; **2** *Eidolon helvum*, **3** *E. pomorphorus walbergii* (after Kingdom, 1971).

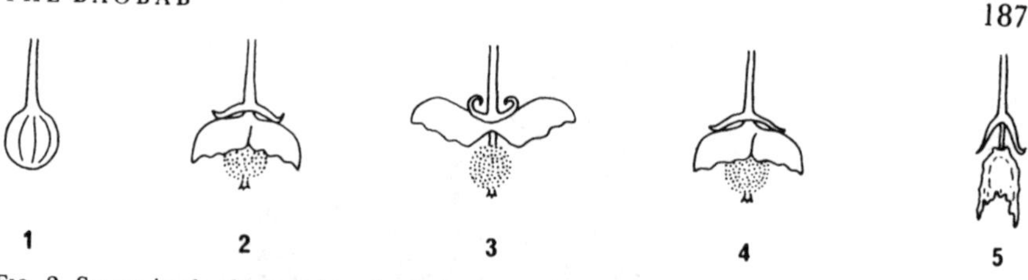

FIG. 2. Stages in the 24 hr life-cycle of the baobab flower (after Breitenbach & Breitenbach, 1974). **1** flower bud just about to burst open; **2** flower just after opening in the evening; **3** following morning showing maximum opening of the flower with sepals and petals reflexed to fully expose the stamens; **4** flower at noon; **5** wilted flower in the evening showing the collapse of the calyx and corolla tube sliding down the staminal tube.

reported small bats visiting the flowers of baobab trees planted at Buitenzorg.

A decade was to pass before similar observations were made in its native continent, and reports of the fruit bat *Eidolon helvum* pollinating the flowers came in from West Africa. The same species was also recorded in Kenya in 1965, from where two other fruit bats *(Rousettus aegyptiacus* and *Epomorphus wahlbergii)* were later reported (Map 2B).

The distribution of these fruit bats is shown in Map 2; it is perhaps significant that the distribution also coincides with that of the baobab. However, the pollen is so lightly held by the anthers and the stigma so receptive that wind pollination may also be possible.

At night the bats swoop down and attack the flowers, albeit for a few seconds only, clinging by means of their claws to the petals or staminal tube, which are so rich in tannins that any damage caused by the claws soon oxidises to form conspicuous brownish scars. The bats are believed to seek out the nectar that is secreted on the inner basal area of the sepals; the surface is wrinkled and covered by numerous unicellular protective hairs and shorter secretory hairs. Since some quantities of pollen have also been found in the bat's digestive system, and the bats often alight on the flower head downwards, it is believed that the pollen too may be eaten.

In East Africa the bush baby, *Galago crassicaudatus,* feeds nocturnally on the flowers (Plate 8A); perhaps they also aid pollination. Because of the popular appeal of the bush baby, this idea has attracted a great deal of attention out of all proportion to the rather restricted distribution of the animal (Map 2A) in relation to the baobab (Map 1).

The flowers emit what some describe as a strong carrion smell, which is presumably attractive to the bats; it is also known to attract the bluebottle, *Chrysomyia marginalis* and at least three nocturnal moths, the American Bollworm, *Heliothis armigera,* Red Bollworm, *Diparopsis castanea* and the Spring Bollworm, *Earias biplaga.* Hymenopterous insects also take advantage of the minute opening of the sepals in the late afternoon, before the flowers fully open, to enter the bud and collect pollen grains from the ripe anthers and may thus assist in the pollination.

(e) *Seed dispersal and germination*

When the heavy fruits fall the woody husk often fractures, enabling termites to enter and devour the sweet fruit pulp; by so doing they free the seeds. Some seeds may germinate *in situ* during the rainy season but many of them are either smothered by the surrounding ground vegetation, killed by the shade of the parent tree or destroyed by fire or cultivation. Some of the seeds may be carried further afield by monkeys, squirrels and rats and may even survive the first rainy season, only to be killed by drought during the ensuing dry season. Certainly seed transport by monkeys or baboons offers a reasonable explanation for the occasional presence of a baobab perched on the summit of a hill or rock outcrop. The fruits are certainly eaten by elephants, also by eland and birds; people eating the sweet pulp spit out the seeds and thereby disperse them.

It is possible that the fruits from riverine trees are water-dispersed. Drift seeds recognizable as seeds of some species of *Adansonia* have been washed up on the shores of South Africa as well as on the island of Aldabra; the latter were found to be viable. This is interesting because *Adansonia gregorii* is found in the Kimberley District of West Australia, where it presumably arrived during the Tertiary (see p. 201).

The reports on the germination of the seed are varied. Some were found to germinate in about 3 weeks; others were found to germinate readily after the pulp soaked off in water. But there have been less satisfactory results. In one experiment there was only 8% germination after 189 days following treatment with cold water, 9% with hot water and 7% from the untreated control. In another there was even less: only 1% germination despite trying many of the standard treatments. More disappointing still was the failure to germinate any seed at Kew!

(f) *Association with animals other than pollinators*

The baobab provides food and shelter for a number of insects, reptiles, mammals and birds. Fruit bats eat the young tender leaves and, when fallen, the old dry leaves are eaten by cattle and presumably by other grazing mammals. Elephants also browse; they may do this for the high mucilage content, not for minerals, since they are apparently not interested once the tree is felled. Elephants will also uproot baobabs, strip off the bark and chew the fibrous wood, which they spit out after they have managed to extract all the salts and juices. It has also been suggested that the uprooting of the trees may reflect some form of aggressive behaviour when elephant population pressure becomes too great.

Many animals eat the fruit once the outer shell has withered and broken. Baboons and monkeys break the ripe pods to get at the sweet white pulp within. In Tanzania both elephant and impala eat the fruit; the latter will also eat the fallen flowers.

Many birds frequently roost or nest in the baobab, including the red-winged starlings, swifts and the gregarious buffalo-weavers with their conspicuous nests dangling at the ends of the branches; others, such as the rollers, hornbills, parrots and kingfishers use holes in the trunk. Lovebirds, barn owls and sometimes Wahlberg's eagle also nest in the tree, the latter providing a protective umbrella against potential nest robbers of the other birds.

The baobab also provides shelter for the bush baby and various venomous snakes, including the boomslang. Wild bees, sweet bees, stick insects and many others, including the numerous insect pests noted in the following section also nest in the trees.

Considering the large number of people who profess an interest in the African fauna surprisingly little has been published about the fauna associations of the baobab.

PLATE 5

Forms of the baobab. **A** trunk tapering, branching from low down—in this particular example the trunk has split. Tanzania, Tanga, Jan. 1976. Photograph by P. J. Cribb; **B** bottle-shaped trunk. Tanzania, W of Mombo, Feb. 1976. Note the beehives hanging from the branches. Photograph by C. Grey-Wilson; **C** stout, cylindrical trunk, branching from the top, Malaŵi, near Lake Nkopola Lodge, April 1980, Photograph by C. C. Townsend; **D** trunk short, squat. Transvaal, c. 60 km N of Lydenburg, Oct. 1968. Photograph by Mrs Davidson.

PLATE 6

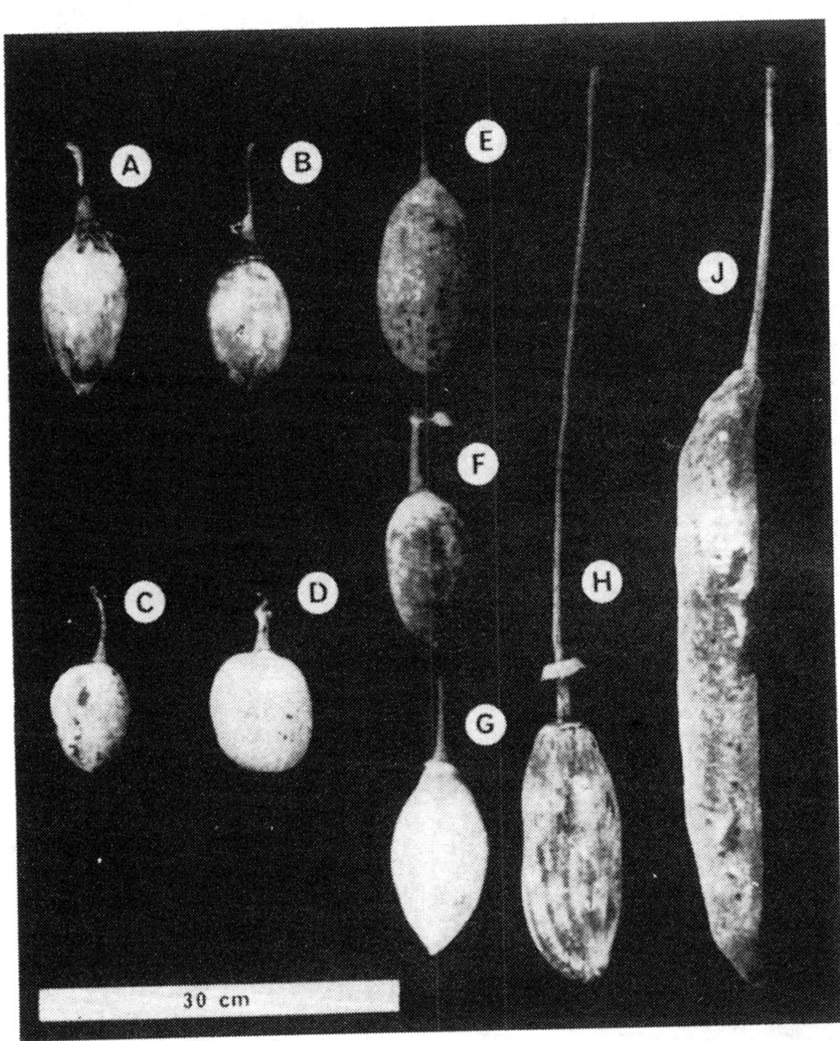

Some of the range of variation found on the fruits of *Adansonia digitata*: **A–D** ± globose fruits with short pedicels from S. Africa, Messina Road on the northern slopes of the Zoutpansberg, near Msequa's Poort, *Leach* 14773 (SRGH); **E** bluntly ovoid fruit from Zimbabwe, Chipinga District, near Sabi Experimental Station, June 1957, *Bates* s.n. (SRGH); **F–G** sharply ovoid fruits from Botswana, Sizora Pan, ± 50 km W of mouth of Nata River, *Drummond & Seagrief* 5244 (SRGH); **H** large ovoid and shallowly sulcate fruit with a long pedicel from Angola, Parc de Cachoneiras, E. of Novo Redondo, *Leach & Cannell* 14612A (SRGH); **J** oblong-cylindrical fruit with a long pedicel from Angola, ± 45 km E of Benguella, *Leach & Cannell* 14605 (SRGH). Photographs by L. C. Leach.

PLATE 7

A The stump of a baobab from Dakar, Senegal, showing the cut surface covered by bark, with a central shoot and several peripheral shoots. The rule represents 1 metre. Photograph by courtesy of I.F.A.N. **B** An avenue of baobabs in a village near the Gambia River showing how the bark is stripped for fibre. (Photograph by Allison Armour, West African Expedition 1927).

PLATE 8

A A bushbaby pollinating a baobab flower. Flash-light photograph by an unknown photographer. **B** A flower early in the morning, Tanzania, W of Mombo, Feb. 1976. Photograph by C. Grey-Wilson.

(g) *Pests and diseases*

Because of the great interest in the alternate hosts for insect pests of cotton and cocoa, the insect fauna of the baobab has been investigated rather more thoroughly than for most native African trees.

They include the cotton bollworms *Heliothis armigera, Diparopsis castanea* and *Earias biplaga*, cotton-stainer bugs such as *Dysdercus fasciatus, D. intermedius, D. nigrofasciatus, D. suberstitiosus, Odontopus exsanguinis, O. sexpunctatus* and *Oxycarenus albipennis* as well as flea beetles, *Podagrica* spp. The baobab is also a host for members of the *Pseudococcoidae*, the mealybugs, which act as vectors for the various virus diseases of cocoa in West Africa as well as the cocoa capsid, *Distantiella theobroma*. Pollarding was formerly practiced in the Sudan Republic for the control of the cotton-stainers (and was also found to stop fruiting for at least two years).

There have been a number of ill-advised attempts to protect both the cotton and cocoa crops by eliminating the baobab, which have later proved to be ineffective because of the presence of other and often more numerous but less conspicuous alternative hosts in the *Bombacaceae, Malvaceae, Sterculiaceae*, etc.

In Ghana an unidentified black beetle is reported to damage and eventually destroy branches by girdling. Also from West Africa there is a report of a longhorn beetle, *Aneleptes trifascicata*, which will attack and kill young trees by girdling the stem.

In the Transvaal the masonga caterpillar or mopane worm, *Gonimbrasia herlina* is said to feed on the leaves.

There are surprisingly few observations relating to fungi. The only macrofungi recorded are those in Saccardo (1898, p. 40); they are *Daldinia concentrica* (Bolt.) Ces. & De Not., *Coriolopsis stumosus* (Fr.) Ryvarden (syn. *Polystictus luteo-olivaceus* Berk. & Br.) and *Trametes socrotana* Cooke. There are only two records for fungal disease, a leafspot in the Sudan Republic due to *Phyllosticta* sp. (Tarr 1955) and a powdery mildew in Tanzania caused by *Leveillula taunica* (Lév.) Arnaud (Wallace & Wallace 1944; Ebbels & Allen 1979).

The fruitful records possible from decaying wood do not appear to have been investigated. Excessive watering of young trees is said to cause stem-rot, but whether this is a fungal or physiological disorder is not recorded.

In West Africa the presence of the baobab of the cocoa mottle-leaf virus and swollen-shoot virus has been thoroughly investigated by a number of researchers. Both Posnette *et. al* (1950) and Tinsley (1955) conclude that the distributional data concerning the original outbreaks of virus infection in the cocoa crops of West Africa indicate that the infection must have been transmitted by mealybugs from forest members of the *Bombacaceae, Sterculiaceae* and *Malvaceae* rather than from naturally infected *Adansonia digitata* in the surrounding savanna.

Although not a serious pest, the mistletoe, *Loranthus mechowii* Engl. has been recorded as a parasite on the baobab in Angola; doubtless other species of *Loranthus* occur elsewhere. Parasitic figs (*Ficus* spp.) have also been seen.

8. Economic importance

Throughout its African distribution the baobab plays a useful and valuable rôle in the economy of the local inhabitants. Practically all parts of the tree can be utilized. It provides food for both man and his livestock, shelter for the living and the dead, clothing, medicine, as well as sundry necessities for hunting, fishing and entertainment. These various uses are discussed in relation to the various parts of the tree, roots, trunk, leaves, flowers and fruit.

(a) *Roots*

In West Africa the roots are reputed to be cooked and eaten, presumably in times of famine. The Temne of Sierra Leone believe that a root-decoction taken with food causes stoutness. The dried powdered root prepared as a mash may be taken by malaria patients perhaps as a tonic.

In East Africa a soluble red dye is obtained from the roots.

In Zambia an infusion of the roots is used to bath babies in order to promote a smooth skin. The root bark is used as string or rope for making fishing nets, socks, mats, etc.

(b) *Hollow trunks*

The hollow trunks of living trees are used in many ways. By far the commonest use is water storage, a practice first recorded by Ibn Batuta from naturally hollowed trees in Mali during the early fourteenth century. Artificially, a hollowed trunk may be carved out in 3–4 days; a medium-sized tree may hold 400 gallons while a large tree could contain 2000 gallons, and water is reputed to remain sweet in them for several years if the hollow is kept well closed. The water can become polluted with soil and organic matter during the process of filling the cavity. In the Sudan this turbidity is traditionally reduced by placing twigs of *Boscia senegalensis* or *Maerua crassifolia* (both of the Caper family, *Capparidaceae*) in the water. Such water storage is frowned upon by medical officers who maintain that it provides breeding sites for mosquitoes.

The opening to the hollow is preferably made just above a branch so that some water may run directly down the branch into the hollow. Whenever possible the people dig out a shallow dip in the ground for rainwater to collect in and then scoop the water up with a bucket and a long rope to the opening in the trunk.

In the Sudan the trees are regarded as personal property that may be inherited or sold and the ownership of the various trees is kept in local government registers. The trees were often the only source of water available during the dry season for both villagers and long-distance travellers; alas, many were deliberately destroyed during the time of the Mahdi to prevent the movement of people.

Hollow trees filled with water in the Northern Frontier Province of Kenya were used by slave and ivory raiders from Ethiopia, enabling them to cross otherwise waterless country.

Early records of European travellers in South Africa noted the use by the Bantu tribes of water stored in natural hollows. A chain of trees across the Kalahari to South-West Africa was used for water storage, with a long-established tradition of death to a traveller who left the bung out of a tree and wasted water. Bushmen are reported to suck water from hollow trunks with grass straws.

In West Africa the hollow trunks may be used as tombs. In Senegal the Sérères make use of hollow baobabs as a place where a body denied burial may be suspended between earth and sky for mummification so that their bodies do not pollute the earth. The people denied normal burial are the griots, a caste which includes poets, musicians, sorcerers, drummers and buffoons.

At Grand Galarques in Senegal a hollow tree with a carved entrance was used as a meeting place. Another tree in Nigeria was used as a prison, and from Mali came the report by Ibn Batuta of the weaver's loom set up in a hollow tree. Elsewhere in West Africa hollowed trees have been used as stables.

In East Africa too the trunks are hollowed out to provide shelter and storage but it is in southern Africa that one finds their most varied use. In Zimbabwe a hollow tree on the Birchenough Bridge road is used as a bus shelter, accommodating 30–40 persons. A tree near Umtali is used by the Roads Department for keeping wheelbarrows and implements in, and another in the Triangle Sugar Estates is fitted with crude ladders and used as a watchtower for cane or veld fires.

There is a famous baobab outside the Commissioner's Office at Katimo Mulilo (17° 30′ S, 24° 20′ E) in the Caprivi Strip which is fitted out with a flush toilet (Plate 4). The story told is that a former Commissioner during the 1939–45 War, a Major Trollip, became so bored with nothing to do that he instructed his staff to install it.

Near Leydsdorp in the Transvaal the hollow trunk of a baobab is used as a cool room for a tin shack beneath the canopy of the tree that is used as a bar. An opening in the top of the trunk creates a cooling draught.

G. L. Guy reports a baobab in the Transvaal being used as a dairy and yet another in Botswana being used as a dwelling.

Since the buffalo-weaver birds in Northern Transvaal are reputed to always rest on the western side of the baobab, the tree can serve in lieu of a compass!

(c) *Bark*

The bark fibres are commonly completely stripped from the lower trunk (Plate 7B). Despite this injury (normally fatal to other trees) baobabs survive and regenerate new bark. The fibre from the inner bark is particularly strong and durable and is widely used for rope, cordage, harness straps, strings for musical instruments, baskets, nets, snares, fishing lines, fibre cloth, etc. In both Senegal and Ethiopia the fibres are woven into waterproof hats that may also serve as drinking vessels!

The dried bark was once exported to Europe for the manufacture of packing paper, and since 1848 it has also been imported into Europe under the name of *cortex cael cedra* and has been used in fevers and as a substitute for cinchona bark. Its benefit as a febrifuge, however, has not been detected in experimental malaria treatments, although it is both diaphoretic and antiperiodic. The bark is certainly used for the treatment of fever in Nigeria.

The bark contains a white, semi-fluid gum which oozes from wounds. The gum is odourless, tasteless, acidic and insoluble and is used for cleaning sores.

The bark tastes bitter; there are uncorroborated accounts of it being eaten in Senegal. As for whether the bark contains alkaloids there are conflicting accounts. One states that none are present, another is inconclusive, a third states that it contains the alkanoid *adansonin* which has a strophanthus-like action—yet in East Africa the bark is used as an antidote to strophanthus poisoning! In Malaŵi the flesh of an animal killed by a poisoned arrow has the juice of the baobab bark poured into the arrow wound in order to neutralize the poison before the meat is eaten.

In some countries the bark is used for tanning. In Congo-Brazzaville a bark decoction is used to bathe rickety children and in Tanzania as a mouthwash for toothache. The ash from the bark and fruit boiled in oil is used as a soap.

(d) *Wood*

The wood is light, 53 lb/cu.ft wet, c. 13 lb/cu.ft air-dried. It is spongy and easily attacked by fungi. If left in water it disintegrates in about 2 months, producing long fibres that could be used for packing.

The wood is not easy to cut; axes bounce off rather than cut the spongy tissue, the force of the blow being absorbed by the elasticity of the parenchyma cells. It makes poor firewood and charcoal, and is not suitable for cutting into planks. The wood can be used for making wide, light canoes, wooden platters, trays, floats for fishing nets, etc.

The wood pulp has been considered suitable for both wrapping and writing paper but attempts to exploit the wood commercially for paper making have so far failed because of the cost of extracting the moisture from the turgid tissues.

(e) *Leaves*

The leaves, especially the young leaves, are popular as a spinach; sometimes they are dried and powdered and made into soups or sauces. The trees may be pollarded in order to encourage an abundance of young leaves (pollarding is also carried out on hollow trees used for water storage in order to prevent them becoming top-heavy and falling over). The leaves are also browsed by stock, being an especially valuable fodder for horses.

The fresh leaves are rich in vitamin C as well as containing uronic acids, rhamnose and other sugars, tannins, potassium tartrate, catechins, etc.

The leaves are used medicinally as a diaphoretic, an expectorant, and as a prophylactic against fever, to check excessive perspiration, and as an astringent. The leaves also have hyposensitive and antihistamine properties, being used to treat kidney and bladder diseases, asthma, general fatigue, diarrhoea, inflammations, insect bites, guinea worm, etc. David Livingstone treated indolent sores with poultices of powdered baobab leaf; the success of the treatment may have been because the ulcers were of dietetic origin.

(f) *Flowers*

The only recorded economic use of the flower is the mixing of the pollen with water to make glue.

(g) *Fruit*

Why the fruit was brought to the ancient Cairo markets is not certain. In the late sixteenth century, the pith was regarded as a substitute for *terra lemnia*, an astringent medicinal earth from the Isle of Lemnos in the Aegean, and as such imported into Europe.

The husks of the fruit may be used as dishes or fashioned into vessels or snuff boxes; they are even used as fishing floats. They can be used as fuel and provide (like the bark) a potash-rich ash suitable for soap-making. The powdered husk or peduncle may be smoked as a substitute for tobacco while the fibres lining the husk are used as a decoction to treat amenorrhoea.

The acid pith, which is rich in ascorbic acid is used as a substitute for cream of tartar in baking. It may also be ground and made into gruel or prepared as a refreshing drink. The pith, though dry at first, has a pleasant wine-gum flavour once moistened in the mouth—this is much appreciated by children. The pectin content of the fruit is low and of poorer quality than commercial pectin. Medicinally the pith may be used as a febrifuge and as an anti-dysenteric, and in the treatment of smallpox and measles as an eye instillation. The pulp is also used to curdle milk, and a decoction used to coagulate the latex of *Landolphia heudelotii*. The pulp is also used for smoking fish; the smoke's acrid smell also drives away insects troublesome to stock.

Both the pith and the seeds, like the bark, seem to contain an antidote to strophanthus poisoning and are carried by a special member of the Shangaan hunting party.

An emulsion of the pith is used by the Fulani herdsmen to adulterate milk, a popular drink with the Kaura farmers whom the Fulani supply.

The seeds have a relatively thick shell which is not readily separated from the edible kernel; this limits their usefulness. The seeds may be eaten fresh or dry, either sucked, ground and used to flavour soups, or roasted to provide a substitute for coffee. The shoot and roots of germinating seeds and seedlings are also eaten. An oil may be obtained by distillation of the seeds, which is used in Senegal on gala occasions. The crushed, roasted seeds are applied as a paste to diseased teeth and gums. When burnt the seeds, like the husks, provide a potash-rich salt used for soap-making.

The seeds are sometimes used to adulterate groundnuts. The pulp and seeds have a feeding value similar to many local leguminous pods and are recommended for feeding to stock late in the dry season when grazing is poor.

9. Cultural importance

It is hardly surprising that many stories relating to the baobab have grown up, considering the grotesque appearance of the tree!

It is widely believed that God planted the tree upside down. Some have elaborated the story: the baobab was first planted in the Congo basin, where it complained of the excessive dampness, so God planted it on the 'Mountain of the Moon', Ruwenzori. Again the tree was unhappy and God, angered by its constant wailing, plucked it out and threw it into a dry area in Africa where it landed upside down.

The Bushmen from southern Africa tell a different story: when the Great Spirit gave trees to the first man, he also gave one to each animal; the hyaena was the last animal and received the baobab, the only tree left and he was so upset that he planted it upside down.

One wonders if there are similar legends for the Madagascan baobabs. A recent and somewhat libellous version is that it is the manifestation of a specification for a tree given to the Ministry of Works, Mombasa. There is another delightful belief that the maternal creator of the baobab was frightened by an elephant.

It is also widely held that there are no young baobabs—they spring into being fully grown.

In East Africa, Resa, the Lord of Rain, is said to live in a great baobab that holds up the sky. In Zambia, there is a belief that the baobab starts out as a creeper that eventually coils itself around a tree, engulfs it and turns into a baobab.

In West Africa spirits are believed to congregate at night beneath the baobab; elsewhere it is believed that the flowers are inhabited by spirits and that a lion will devour anyone rash enough to pluck a flower. (Is this why there are so few collections?) In Tanzania and Zambia it is considered inadvisable to suck or eat the seeds while in crocodile country, for the crocodiles will be attracted. On the other hand, another version is that a draught of water in which the seeds have been steeped will act as a protection *against* crocodiles.

In Zimbabwe, baobabs are regarded as 'tagati'—bewitched, so that misfortune would fall on the countryside if they were removed.

Another belief is that a man who drinks an infusion of the bark will become mighty and strong while in the Transvaal it is believed that a baby boy should be washed in water in which the bark has been steeped, but not too much in case the boy becomes obese, nor must his head touch the water lest it swells.

There is a saying in Senegal that 'The Blacks die when the baobab has lost its leaves; it is the turn of the Whites to die when it has regained its leaves' implying that mortality is highest for the native Senegalese during the cold dry season and for the Europeans during the hot wet season.

The use of the hollow trees by the Sérères of Senegal for the mummification of their griots is discussed in the previous chapter.

When clearing the bush in Upper Volta the baobab is left standing as a fetish tree; children of the Ela born under the sign of this tree (*Kukula*, Ela), are given the patronymic *Kukula* for boys or *ekulu* for girls. In northern Nigeria primitive tribes cut symbols in the bark. In southern Nigeria certain trees may be worshipped as a fertility symbol.

In East Africa the distribution of the baobab is reputed to be associated to some extent with trees planted on the graves of Arab traders. It is the totem of the Sebola clan of the Twamumba tribe in Zimbabwe and in the Limpopo Valley women's breasts are depicted in rock paintings as baobab pods.

The baobab is the tree emblem for the Republic of Congo and is widely portrayed on stamps.

10. TREE CLEARING

Engineering schemes involving the clearance of land for dam construction such as Kariba on the Zambesi and Roseires on the Blue Nile or for other purposes has sometimes meant the destruction of many baobabs.

In Senegal the injection of an arsenic insecticide and caustic soda into the trunk at the start of the growing season proved successful, with as much as 80% of the trees falling of their own accord, while the process of cutting down the remainder was very much easier.

In the Sudan trials using two crawler tractors with front-mounted rakes and pulling a heavy anchor chain and one crawler in the rear with a push-over boom proved effective for all but the largest trees; the latter refused to be uprooted, pushed over or destroyed by explosives and were left standing.

In East Africa it was found that a hawser could be used to cut through the trunk—rather like cheesewire. The tree would remain standing and would then have to be pushed over!

11. Conservation

I have already pointed out (p. 189) the futility of eradicating the baobab in order to control pests and diseases of cotton and cocoa, and on p. 192 that attempts had been made to manufacture paper from the wood.

On p. 180 there is the suggestion that the baobab has disappeared from a number of localities in Somalia. It is also being destroyed by elephants in a number of game reserves. Although it is an extremely large and conspicuous tree it is not particularly abundant. The fact that it is extremely long-lived and yet not particularly abundant seems to imply that the natural regeneration is rather poor. Regeneration is taking place in the higher rainfall areas, not in the more arid regions, which suggests a slowly contracting distribution.

The conservation of any plant species in the wild is never easy, especially in Africa where there is an understandable pressure by an expanding population for more land for cultivation. However, the baobab is a conspicuous and attractive feature of the landscape within many of the game reserves of southern and tropical Africa; it is to be hoped that good reserve management will ensure that these trees survive.

They can also be an interesting amenity tree for urban parks. Some effort should be made to ensure that at least some of the more grotesque forms found in the wild are perpetuated by the planting of their seed in parks and other public places where they are likely to be allowed to grow for centuries to come.

12. Bibliographic key to part I

References substantiating statements made in the text are given here, grouped under the sections, subsections and paragraphs to which they are relevant. Full details are given on p. 202 in the collected Reference section.

(2a) § 1. Ancient Egypt: Bonnet 1895, Loret & Poisson 1895, Beauverie 1935, Guy 1971.
§ 2. Harkhuf: Loret & Poisson 1895. Sudan: Schweinfurth 1868.
§ 3. Origin of name: Mauny 1951, Drar 1970, Leriche 1954, Nicolas 1955, Société Horticole d'Alexandre 1901; Schweinfurth 1912 did not mention *habhab*, previously listed in Ascherson & Schweinfurth 1887.
§ 4. Application of 'baobab': Adanson 1757 & 1763–4, Mauny 1951.
§ 5. teidoûm: Leriche 1954.
§ 6. Clusius' illustration: Gerber 1895 deplored the suggestion of numerous 'funicles' to the seed—they were in fact an accurate portrayal of the fruit fibres.

(2b) § 1. Ibn Batuta's travels: Ibn Batuta 1829 in which water storage mentioned.
§ 2. Adanson must have provided the information that the leaves were digitate, subsequently communicated to Linnaeus by Jussieu.
§ 3. Cordoso: Ficalho 1884. Cada-Mosto: Cooke 1870, Gerber 1895.
§ 1. Axelrod 1970 emend. Willis 1973.
§ 2. *Ochroma pyramidale* (Cav. ex Lam.) Urb. (*Bombax pyramidale* Cav. ex Lam. April 1788; *Ochroma lagopus* Sw. Jun. Jul. 1788.)

(5) § 1. Gabon: Walker 1953. Central African Republic: Guigonis pers. comm. 1968. São Tomé, etc.: Exell 1944.

(5a) § 2. Sudan streamsides: Keay 1949. Farmlands: Keay 1953, Howard 1976.
§ 3. Association of baobabs with villages: Chevalier 1906, Aubréville 1950, Adam 1962.
§ 4. Kuka, origin of name: Nachtigal, transl. Fischer & Fischer in press.
§ 5. Zinder area: Unwin 1920.
§ 6. French W. Africa: Adam 1962.
§ 7. Ghana, Togo & Dahomey: Unwin 1920, Irwin 1930, Aubréville 1950. Congo: Aubréville 1950. Bas-Zaïre: White & Werger 1978.

(5b) § 1. Chad: Chevalier 1906, Aubréville 1950.
§ 2. Zalingei, Sudan: Wickens 1977.
§ 6. Disappearance from S Somalia: J. B. Gillett pers. comm. 1979.

(5c) § 1. Kenya coastal: Lind & Mollison 1974, Fenner 1980. Tanzania & Mozambique: Moll & White 1978. Congo–Wembere divide: Gillman 1949.

(5d) § 1. Mopane woodlands: Werger & Coetze 1978.
§ 3. Zoutpansberg mts: Galpin pers. comm. 1969.
§ 4. Optimum conditions for baobabs: Palmer & Pitman 1972.

(6) § 1. Dhofar: Radcliffe-Smith 1979 & verbal info. 1979. Samsara: J. R. I. Wood verbal info. 1979.
§ 2. Madagascan introduction: Perrier & Hochreutiner 1955.
§ 3. Indian distribution: Burton-Page 1969. Association with shrines and religious ceremonies: Vaid 1978a–c. Non-utilization: Vaid & Vaid 1978. Not in ancient Floras, no Sanskrit name: Burton-Page 1969. Identification of baobab with *Kalpa-vriksha*: Vaid 1964, 1978a; and disagreement: Maheshwari 1971. Identification of banyan with *Kalpa-vriksha*: Sinha 1979. Introduction to Ceylon: Robyns 1970.
§ 4. Gardens of Cairo: Ascherson & Schweinfurth 1887. Mauritius: Thompson 1816. Florida: Fairchild 1931.

(7a) § 1. Sixth leaf digitate: Lubbock 1892. Irregular teeth after 10–12 leaves: Perrier 1952. Simple leaves long-persistent: Guy 1971. Leafy period mid-October to mid-February: Fenner 1980 who also discusses moisture uptake and loss.
§ 3. Three stages of growth: Galpin pers. comm. 1969.
§ 4. Short massive branched trunk: Schimper 1903. Sapling trunks coalescing: Woodruff 1969.
§ 5. Branches of young trees spread and droop with age: Hollis 1963; between 30–40 years: Guy 1971.
§ 6. Migration of regenerative bud: Stahel 1977.
§ 7. Turnip-like taproot: Wild 1961.

(7b) § 1. Difficulties in girth measurement: Adam 1963. Girth fluctuation after rain: Galpin pers. comm. 1969. Volume increment more reliable and difficulty in counting annual rings by borings, large diameters: Guy 1970.
§ 2. Adanson's calculations: Adanson 1759. disappearance of Adanson's 2 trees from Magdalene Is.: Adam 1962. Livingstone's wrath: Guy 1971. Livingstone's own measurements: Livingstone 1857.
§ 3. Guy's researches: Guy 1970 & 1971.
§ 4. Agreement of annual increment with annual ring width: Swart 1963. Largest trees not always oldest: Lafont 1942.

(7c) § 1. W African flowering: Dalziel 1937. S African flowering: Palmer & Pitman 1972. Zimbabwe flowering: Pardy 1953. Sudan: Hunting Technical Services 1964.
§ 2. Sequence of flower opening: Harroveld-Lako 1926, Jaeger 1950 & 1961, Adam 1962, Breitenbach & Breitenbach 1974.

(7d) § 1. Bat pollination confirmed: Pijl 1936.
§ 2. For W Africa: Jaeger 1945 & 1950, Baker & Harris 1959, for Kenya: Coe & Isaac 1965; two other bats: Start 1972, Koch 1972.
§ 3. Pollen only lightly held: Jaeger 1945.
§ 4. Pollen perhaps eaten: Jaeger 1961.
§ 5. Bush-baby: Coe & Isaac 1965.
§ 6. Bollworms: Guy 1969 & 1971, Strover pers. comm. 1970. Hymenoptera: Jaeger 1961.

(7e) § 1. Fruit fracturing: Chevalier 1906. Fruit eaten by eland: Ridley 1930, Owen 1974; birds: Guy 1971; people: Chevalier 1906, Hobley 1922, Ridley 1930, Guy 1971.
§ 2. Drift seeds, S Africa: Muir 1937; Aldabra: Wickens 1979, & their viability: Hnatiuk verbal info. 1980.
§ 3. Germination: 3-week: Vauquelin 1822. Easier after soaking: Batty pers. comm. 1970. Less satisfactory results: Anon. 1955. Standard treatment 1%: Astle pers. comm. 1969.

(7f) § 1. Bats eat young leaves: Sweeney 1969. Lack of elephant interest after felling: Astle pers. comm. 1969. Elephant damage: Bax & Sheldrick 1963, Astle pers. comm. 1969, Strover pers. comm. 1970, Owen 1974. Aggression: S. K. Sikes quoted in Owen 1974.
§ 2. Fruits eaten if shell broken: French 1944. Impala eat flowers: Lamprey 1963, Owen 1974.
§ 3. Use of holes in trunk: Strover pers. comm. 1970; Owen 1974. Eagle: Owen 1974.
§ 4. Insects nest in baobabs: Strover pers. comm. 1970; Owen 1974.

(7g) § 2. Insects: Crowther 1948, Sweeney 1969, Strover pers. comm. 1970, Kranz *et al.* 1977. Pollarding: Crowther 1948. Mealybugs: Kranz *et al.* 1977.
§ 4. Girdling damage: Roberts 1961, O. A. Atindale & Obeng-Darko pers. comm. 1969.
§ 5. De Winter *et al.* 1966.
§ 7. Stemrot after excessive watering: Breitenbach 1965.
§ 8. Virus: Posnette *et al.* 1950, Tinsley 1955, Deighton & Tinsley 1958, Dale & Attafuah 1958, Attafuah & Tinsley 1958, Legg & Bannay 1967, Kenton & Legg 1971; distribution data: Posnette *et al.* 1950, Tinsley 1955.

§ 9. Angola mistletoe: Hiern 1896. Parasitic figs: Galpin pers. comm. 1968.

(8) Uses follow earlier paper: Wickens 1979.

(8a) § 1. Malaria: Burkill ined.

§ 2. Burkill ined.

§ 3. Zambia, fibres: D. B. Fanshawe pers. comm. 1969.

(8b) § 1. Water storage: Ibn Batuta 1829; Drar 1970; Nachtigal 1971. Turbidity reduced: Jahn 1979. Mosquitoes: Burkill ined.

§ 3. Ownership, Sudan: Nachtigal 1971. Mahdi: Owen 1970.

§ 4. E. A. Galpin pers. comm. 1969.

§ 5. Water in natural hollows: Story 1964. Chain of trees in Kalahari: Mogg 1950. Grass straws: Story 1958.

§ 6. Cooke 1870, Grisard 1891, Mauny 1955, Drar 1970, Owen 1970, Guy 1971; Armstrong 1977; Burkill ined.

§ 7. Meeting place: Cooke 1870. Prison: Owen 1970. Weaver: Ibn Batuta 1929. Stables: Guy 1971.

§ 8. E. C. Strover pers. comm. 1970, Guy 1971, Burkill ined. Watchtower: E. C. Strover pers. comm. 1970.

§ 9. Fanshawe pers. comm. 1969; Killick pers. comm. 1969; Strover pers. comm. 1970; Guy 1971.

§10. Bar: Koeleman 1972. Dairy: Guy 1971.

§11. E. A. Galpin pers. comm. 1969.

(8c) § 1. Fibres: Giscard 1891, Williamson 1955, Sabri 1968, Sweeney 1969, Woodruff 1969, Drar 1970. Hats: Grosard 1891, De Wildeman 1903.

§ 2. Paper: Dalziell 1937, Burkill ined. Substitute for Cinchona: Watt & Breyer-Brandwijk 1962. Diaphoretic & Antiperiodic: Burkill ined. Against fever: Oliver 1959.

§ 3. Gum properties: Watts 1889; Uses: Burkill ined.

§ 4. Alkaloids absent: Loustalot & Pagan 1949. Inconclusive: Burkill ined. Adansonin present: Watt & Breyer-Brandwijk 1962.

§ 5. Mouthwash: Burkill ined. Ash: Rock 1861, Watt 1889.

(8d) § 1. Weight: Pardy 1953. Fibres after soaking: Sahni 1968.

§ 3. Wood pulp use: Seabra 1948, Carvalho 1953. Turgidity: Mogg 1950.

(8e) § 1. Edibility: Williamson 1955, Nicol 1957, Sahni 1968, Woodruff 1969, Owen 1970, E. C. Strover pers. comm. 1970. Horsefodder: Woodruff 1969, Burkill ined.

§ 2. Vitamin C: Carr 1958. Other compounds: Watt & Breyer-Brandwijk 1932.

§ 3. Medicinal: Watt & Breyer-Brandwijk 1962, Carr 1959, Woodruff 1969, Beltrame 1974, Burkill ined. Ulcers: Carr 1959.

(8f) Woodruff 1969.

(8g) § 1. Ancient Egypt: Loret & Poisson 1895; Beauverie 1935.

§ 2. Dishes: Ibn Batuta 1829. Snuffboxes and tobacco substitute: Burkill ined. Floats; Watt 1889.

§ 3. Ascorbic acid: Carr 1958. Cream of Tartar substitute: Mogg 1950, Williamson 1955. Pectin content: Nour *et al.* 1980. Eye instillation: Burkill ined. Latex coagulation: Chevalier 1906.

§ 4. E. C. Strover pers. comm. 1970.

§ 5. Nicol 1957.

§ 6. Ash for soap: Burkill ined.

§ 7. Groundnut adulterant: Chevalier 1923. Stock feed: Pelly 1913, Greene 1932, French 1944.

(9) § 2. Owen 1970, Palmer & Pitman 1972, Armstrong 1977, Palgrave 1977. Complaining: Vaid 1978a.

§ 3. Owen 1970, Guy 1971, Armstrong 1977.

§ 4. Elephant: Palmer & Pitman 1972.

§ 6. Resa: Wright & Kerfoot 1966, Guy 1971. Zambia: Hughes 1933.

§ 7. W Africa: Owen 1970. Danger of picking flowers: Woodruff 1969, Palmer & Pitman 1972, Palgrave 1977. Attracts crocodiles: Woodruff 1969, Guy 1971. Protection against crocodiles: Palgrave 1977.

§ 8. E. C. Strover pers. comm. 1969.

§ 9. Bark infusion: Palgrave 1977, Palmer & Pitman 1972.

§10. Chevalier 1906.

§12. Symbols: Burkill ined. Owen 1970.

§13. Arab graves: Woodruff 1969. Sebola totem: Wright & Kerfoot 1966. Limpopo: Owen 1970, Palmer & Pitman 1972, Burkill ined.

§14. Emblem: Asch 1968. Stamps: Robyns 1972.

(10) § 2. Poisoning: Anon. 1963.

§ 3. Too tough: Schoenwald 1969.

§ 4. Hawsers: Hobley 1922.

PART II

13. Nomenclature and Synonymy

Adansonia digitata *L.*, Sp. Pl. ed. 1: 1190 (1753) & Demonstr. Pl. Hoejer.: 27 (1753); nom. inval., gen. non descr.; Syst. Nat. ed. 10, 2: 1144 (1759). Type: Senegal, *Adanson* s.n. (holotype, LINN).

A. bahobab L., Sp. Pl. ed. 2: 960 (1763), nom. illegit., superfl.
Ophelus sitularius Lour., Fl. Cochin.: 412 (1790). Type: E Africa, not located.
Adansonia baobab Gaertn., Fruct. 2: 253, t. 135 (1791). Type: not located.
A. situla (Lour.) Spreng., Syst. 3: 124 (1826).
A. integrifolia Raf., Sylva Tell.: 149 (1838), nom. superfl., based on *Ophelus sitularius*.
Boababus digitata (L.) Kuntze, Rev. Gen. 1: 66 (1891).
Adansonia sphaerocarpa A. Chev., in Actes Congo Int. Bot. 1900: 271 (1901). Type: Guinée, Timbo, Fouta Djalon, *Chevalier* 12424; Niger, *Chevalier* s.n. (carp.) (syntypes P).
A. digitata L. var. *congolensis* A. Chev. in Bull. Soc. Bot. Fr. 53: 493 (1906). Type: São Tomé, *Chevalier* s.n. (carp.) (holotype, P).
A. sulcata A. Chev. in Bull. Soc. Bot. Fr. 53: 494, pl. 7, figs. 5 & 6 (1906). Type: Brazzaville, *Chevalier* 4230, *Chevalier* s.n. (carp.); *Prince Roland Bonaparte* s.n. (syntypes P).
A. somalensis Chiov., Fl. Somala 8: 30, fig. 10 (1932) in obs.

14. Specific Variation

There is considerable variation within the species which might merit taxonomic recognition, although without adequate material it would be foolish to consider the matter further. Lucas (1971) has already noted the surprising paucity of herbarium representation of such a well-known tree.

Chevalier (1906) considered the shape of the fruit to be an important diagnostic character although his ideas were not generally accepted. There may be some truth in his claim since the fruits in West Africa are larger and are borne on longer pedicels than those from southern and eastern Africa. The West African trees, which are somewhat geographically isolated from the remainder, attain their extreme forms in Angola. Thus, *Leach & Cannell* 14605 collected c. 45 km east of Benguela in southern Angola has oblong-cylindrical fruits 54 × 7·5–8 cm borne on an incomplete pedicel 26 cm long, and from the same area but from a separate population *Leach & Cannell* 14612A represents an oblong-ellipsoid and shallowy sulcate fruit c. 23 × 9·5 cm with a pedicel 57 cm long (Plate 3). Both specimens are at SRGH (Leach, pers. comm. 1971).

We need to know much more about the relation of pedicel length to flower and fruit and the possible range of variation in fruit size and shape both on individual tres and within local populations. This information needs to be correlated with other possible taxonomic characters such as leaf indumentum, trunk form (photograph if possible), etc. The cytology certainly needs further study, with chromosome counts taken from throughout the entire distribution area. This may, or may not, support the writer's tentative conclusion that West African populations might eventually be proved to be taxonomically separable from those of eastern and southern Africa. The chromosome counts shown in Table 1 are certainly suggestive and demonstrate the necessity for further studies.

15. Cytology

Chromosome counts have been carried out for 7 of the 9 species currently recognized and these are shown in Table 1 below.

Table 1. Chromosome counts in *Adansonia* spp.

Species	*2n*	*Reference*	*Distribution*
A. alba Jumell & H. Perr.	?		Madagascar
A. suarezensis H. Perr.	?		Madagascar
A. za Baill.	48	Miège (1974)	Madagascar
A. grandidieri Baill.	60–64	Miège (1974)	Madagascar
A. grandidieri Baill.	88	Baker & Baker (1968)	Madagascar
A. fony Baill.	72	Miège (1974)	Madagascar
A. fony Baill.	88	Baker & Baker (1968)	Madagascar
A. madagascariensis Baill.	80–84	Miège (1974)	Madagascar
A. perrieri Capuron	96	Miège (1974)	Madagascar
A. gregorii F. Muell.	72	Baker & Baker (1968)	W Australia
A. gregorii F. Muell.	96	Miège & Burdet (1968)	W Australia
A. digitata L.	96	Riley (1960)	S Africa
A. digitata L.	128	Schröder pers. comm. 1980	S Africa
A. digitata L.	144	Miège (1960); Miège & Burdet (1968) and Baker & Baker (1968)	W Africa

The counts for the species show an interesting range in polyploidy, based on x = ?12 or ?8 (Miège 1974). Not only are the greatest number of species to be found in Madagascar, the chromosome counts also show it to be the centre of speciation with the lowest count of 2n = 48 with increasing polyploidy to 2n = 96. There are a number of discrepancies in the counts by different authors. To some extent these can be accounted for by the difficulties in counting the very small chromosomes and also the difficulties in identifying correctly rather poorly described polymorphic species. This does not explain the noticeable differences within *A. gregorii* and *A. digitata*. I can offer no explanation for the former but for *A. digitata* it should be noted that the counts of Miège and Baker & Baker are from seed collected in West Africa while that of Riley and Schröder are from South Africa. In view of the taxonomic variation noted previously and the interrupted distribution between West Africa and the Sudan southwards (Map 1), it would be interesting to have many further counts, particularly from eastern Africa.

The difficulties in obtaining well-stained chromosomes from the root tips are briefly discussed by Riley (1960). Baker & Baker (1968) describe the persistence of the nucleolus throughout nuclear divisions, a characteristic feature of cell division in the *Bombacaceae*. The nucleus and mitosis in *Adansonia digitata* are described by Miège & Burdet (1967), while Miège (1975) also reported that the increase in protein spectra observed by electrophoresis of seed proteins in species of *Adansonia* was not related to the increase in chromosome number.

16. Palynology

The pollen of *Adansonia digitata* and *A. grandidieri* is briefly described by Erdtman (1952); other species from Madagascar are described by Prestling

(1962). Fuchs (1966) remarks that pollen from the Australian representatives (some authorities regard *A. muelleri* as distinct from *A. gregorii*) is indistinguishable from that of Madagascan and African species.

The grains of *A. digitata* are illustrated and described in detail by Sharma (1970). They are 3-zonoporate and spinuliferous, thus differing from species in other genera of subtribe *Adansonieae,* which are 3-zonocolporate with various types of exine patterning. Dry pollen was found to be between 58·3–90·63 μm in diameter (Davis & Sukhendu 1976), with up to 30% expansion on wetting.

17. ANATOMY

For a general description of the family see Metcalfe & Chalk (1950). More specialized investigations include observations by Johnson (1961) that in addition to the usual tunica-corpus zonation, the apex of *Adansonia* and other genera of the *Bombacaceae* is characterized by a prominent central zone of metrameristem surrounded by a flanking zone and underlain with maturing pith or a pith rib meristem.

The comparative anatomy of *Adansonia digitata, A. gregorii* and *A. madagascariensis* is discussed by Gerber (1895). More detailed studies of *A. digitata* were undertaken by Braun (1900) and published in a little known work that deserves far greater recognition. It was Braun who recognized the presence of annual rings, an ill-defined feature which some authorities have had difficulty in recognizing.

Young plants do not have a succulent cortex, or any other succulent tissue. As a result inexperienced observers have sometimes been deceived by their non-succulent habit and failed to recognize them. The grossly enlarged trunks of the older trees are composed almost entirely of wood, with water-storing parenchyma particularly abundant (Newton 1974). The density of fresh wood is given as 849 kg/m^3 compared with 208 kg/m^3 when air dry (Pardy 1953), the difference being almost entirely due to the water content. The high water content provided the turgid strength to the tissues.

The surface cells of the leaves have been examined by Imadar & Chohan (1969) who note that they are hypostomatic, the mature stomata being anomocytic, paracytic and anisocytic.

The morphology of the anther is discussed by Heel (1966); the development of the anther, pollen, ovule and embryo-sac by Rao (1954) and the ontogeny of the ovary and style by Heel (1974).

18. GEOLOGICAL HISTORY

The distribution of the species and their cytology indicates a Madagascan centre of origin for the genus. By far the simplest explanation for the present-day distribution is for a direct migration route between Madagascar and Africa and between Madagascar and Australia via India prior to the break-up of the western Gondwanaland land mass. This idea was accepted by Aubréville (1975, 1976).

According to Raven & Axelrod (1974) the break-up took place during the Early Cretaceous, 110 m.y. B.P., with the possibility of a somewhat interrupted subtropical migration route to Australia being possible even in the late Cretaceous, 75 m.y. B.P. However, they argue that the absence of any pollen of the *Bombacaceae* being found earlier than the Palaeocene must imply a late evolution for the family and therefore suggest long-distance dispersal during the Tertiary. The absence of the pollen record cannot and should not be regarded as positive evidence for a late evolution; the chances of pollen preservation, its discovery and identification and the time factor involved are too uncertain.

Gardner (1944) was unable to offer any explanation for the presence of *A. gregorii* in Australia and cited additional examples of Madagascar–Australia affinities, including *Diplopeltis (Sapindaceae)*, now known to be endemic to Australia (George & Erdtman 1969). The two other examples cited were *Rulingia* and *Keraudrenia (Sterculiaceae)*, each with one representative in Madagascar and several species in Australia, the reverse of the *Adansonia* distribution pattern and therefore believed to have migrated under different conditions to those that transported *Adansonia*. Raven & Axelrod (1974) suggest that both *Rulingia* and *Keraudrenia*, amongst others, reached Madagascar by long-distance dispersal around the Indian Ocean. There is still no satisfactory explanation for the presence of *A. gregorii* in Australia. I agree with Perrier & Hochreutiner (1955) that the affinities of *A. gregorii* are with the Madagascan species and not with *A. digitata* as some have suggested.

19. Appendix—Sources, Acknowledgements

The information obtained for this present paper regarding the distribution and ecology of the baobab has been gleaned from the following four major sources.

1. Herbarium specimens, both seen and cited in floristic works.
2. Photographs, both published and unpublished.
3. Proforma returns and correspondence from the Baobab Map Project.
4. Ecological reports, maps, travel literature, etc.

These four sources are complementary, with no single method providing sufficient information to provide an overall picture of the distribution let alone the ecology. Even now there may still be some gaps.

Herbarium specimens and floristic works rarely provided any ecological data, although the latter usually indicated whether the species was native or introduced. It was interesting to note that the distribution data obtained from the regional floras of former Italian Somalia in several instances did not agree with present-day observations resulting from the Baobab Map Project and it is believed that the trees may have been destroyed and that no natural regeneration has taken place.

Professor Walter, formerly of Universität Hohenheim, Stuttgart very kindly presented to Kew all the baobab photographs and correspondence of the late Dr Springer, a former schoolmaster and amateur botanist, which were used to construct his distribution map (Walter 1964). This is an excellent map and this present revision offers very few extensions to the general distribution. It also provided an excellent illustration of the inaccurate picture revealed by relying on herbarium specimens only.

A great deal of information regarding both distribution and ecology was obtained from the proforma returns and accompanying correspondence; the latter was sometimes supported by photographs. I would like to thank the many correspondents who so willingly supported the Baobab Map Project for without their enthusiastic support this paper would not have been possible; even nil returns were of value.

The northern distribution of the baobab in the Sudan has been plotted with remarkable accuracy by Schweinfurth (1868). The 1:250,000 series of maps for the Sudan actually plot individual trees. The regional distribution for Tanzania has been mapped by Gillman (1949), and for South Africa by De Winter *et. al.* (1966). The overall distribution for Africa by Walter (1964) has already been mentioned; Lebrun & Stork (1977) provide further references. It is not proposed to cite the very extensive travel literature examined to obtain further localities since much of the information so obtained was already recorded from other sources; the ecological literature, however, is cited.

20. References

Adam, J. G. (1962). Le baobab (*Adansonia digitata* L.). Notes Africaines No. 94: 33–44.

—— (1963). Le plus gros baobab de Sénégal n'est plus celui de Dakar. Notes Africaines No. 98: 50–53.

Adanson, M. (1757). Histoire naturelle du Sénégal. Paris.

—— (1763–4). Familles des Plantes. 2 vols. Paris.

—— (1771). Description d'un arbre d'un genre nouveau appelé Baobab observé du Sénégal. Hist. Acad. Roy. Sci. 1791: 77–85, 218–243.

Alpino, P. (1592). De plantis aegypti liber. Venice.

Anon. (1955). Investigation on seeds. For Res. India 1950–51, Pt. 1: 18–19.

—— (1963). [Destruction of Baobab] Gazeta Agric. Angola 7: 522–523 (Port.). Forest Abs. 25: 247 (1964).

Armstrong, P. (1977). Baobabs remnant of Gondwanaland? New Scientist 73: 212–213.

Asch, J. (1968). Botanical emblems of the Nations. Garden Journ. N. York 18, 2: 55–57.

Ascherson, P. & Schweinfurth, G. (1887). Illustration de la Flore d'Egypte. Cairo.

Astle, W. L. (1969). Department of Wildlife, Chipata, Fort Jameson, Zambia. pers. comm.

Attafuah, A. & Tinsley, T. W. (1958). Virus diseases of *Adansonia digitata* (Bombacaceae) and their relation to cacao in Ghana. Ann. Appl. Biol. 46: 20–29.

Aubréville, A. (1950). Flore forestière soudano-guinéenne. Paris.

—— (1975). Essais de géophylétique des Bombacacées. Adansonia II, 15: 57–64.

—— (1976). Madagascar au sein de la Pangée. Adansonia II, 15: 295–305.

Axelrod, D. I. (1970). Mesozoic palaeogeography and early angiosperm history. Bot. Rev. 36: 277–319.

Baker, H. G. & Baker, I. (1968). Chromosome numbers in the Bombacaceae. Bot. Gaz. 129: 294–296.

—— & Harris, B. J. (1959). Bat pollination of the silk cotton tree *Ceiba pentandra* (L.) Gaertn. *(sensu lato)*. Ghana Journ. W. Afr. Sci. Assoc. 5: 1–9.

Batty, Mrs M. (1970). TANESCO, Dar es Salaam, Tanzania, pers. comm.

Bauhin, C. (1671). Pinax theatri botanici, ed. 2, Basel.

Bauhin, J. & Cherler, J. H. & Chabrey, D. (1650–51). Historia plantarum universalis. Yverdon.

Bax, P. N. & Sheldrick, P. L. W. (1963). Some preliminary observations on the food of Elephant in the Tsavo Royal National Park (East) of Kenya. E. Afr. Wildlife Journ. 1: 40–43.

Beauverie, M.-A. (1935). Description illustrée des végétaux antiques de Musée Egyptien du Louvre. Bull. Inst. Fr. Arch. Orient. 35: 115–151.

Bertrame, G. (1974). Up the Blue Nile Valley from Sinnar to Bani Shanqul and back, 1854–1855. In E. Toniolo & R. Hill (Eds.). The opening of the Nile basin: 219–248. London.

Bonnet, E. (1895). Le piante egiziane de Museo Reale di Torino. Nuov. Giorn. Bot. Ital. n.s. 2: 21–28.

Braan, K. (1900). Beiträge zur Anatomie der *Adansonia digitata* L. Basel.

Breitenbach, F. von (1965). The Indigenous Trees of Southern Africa. Vol. IV. Pretoria.

—— & Breitenbach, J. von (1974). Baobab flower. Trees in S. Afr. 26: 10–15.

Burkill, H. M. ined. The useful plants of West Tropical Africa.

Burton-Page, J. (1969). The problem of the introduction of *Adansonia digitata* into India. In P. J. Ucko & G. W. Dimbleday (Eds.). The Domestication and Exploitation of Plants and Animals: 331–335. London.

Carr, W. R. (1955). Ascorbic acid content of baobab fruit. Nature 176: 173.

—— (1958). The Baobab tree: a good source of ascorbic acid. Central Afr. J. Med. 4: 372–374.

Carvalho, J. da Silva (1953). Alguns ensaios para o aprovertamente em cellulose dum material fibroso (*Adansonia digitata* L.).? Publ. Serv. flor. agric. Portugal 20(2): 173–196 [For. Abstr. 18: 574 (1957)].

Chevalier, A. (1906). Les baobabs (*Adansonia*) de l'Afrique continentale. Bull. Soc. Bot. Fr. 53: 480–496.

—— (1923). Sur l'adultération des arachides du Sénégal par les graines du baobab. Bull. Mat. Grasses Inst. Colon. Marseille 1923: 402–403.

Clusius [L'Eseluse] J. C. (1605). Exoticum libri decem. Leiden.

Coe, H. J. & Isaac, F. M. (1965). Pollination of the baobab (*Adansonia digitata* L.) by the lesser bush baby (*Galago crassicaudatus* E. Geoffroy). E. Afr. Wildlife Journ. 3: 123–124.

Cooke, M. C. (1870). Baobab, *Adansonia digitata* L. Pharm. Journ. III, 1: 64.

Crowther, F. (1948). A review of experimental work. In J. D. Tothill (Ed.). Agriculture in the Sudan: 439–592. London.

Dale, W. T. & Attafuah, A. (1958). The host range of cocoa virus. Ann. Rep. W. Afr. Cocoa Res. Inst. 1956/57: 20–24.

Dalziel, J. M. (1937). The Useful Plants of West Tropical Africa. London.

Dapper, O. (1686). Description de l'Afrique. Amsterdam.

Davis, T. A. & Sukhendu, S. G. (1976). Morphology of *Adansonia digitata* L. Adansonia II, 15: 471–479.

Deighton, F. C. & Tinsley, T. W. (1958). Notes on some plant virus diseases in Ghana and Sierra Leone. Journ. W. Afr. Sci. Assoc. 4: 4–8.

De Wildeman, E. (1903). Notices sur des plantes utiles ou intéressantes de la Flore du Congo, vol. 1.

De Winter, B., De Winter, M. & Killick, J. B. (1966). Sixty-six Transvaal Trees. Pretoria.

Drar, M. (1970). A botanical expedition to the Sudan in 1938. Ed. V. Täckholm. Cairo Univ. Herb. Publ. No. 3.
Ebbels, D. L. & Allen, D. J. (1979). A supplementary and annotated list of plant diseases, pathogens and associated fungi in Tanzania. Comm. Agric. Bureau Phytopath. Paper No. 22.
Erdtman, G. (1952). Pollen morphology and plant taxonomy. Stockholm.
Exell, A. W. (1944). Catalogue of the vascular plants of S. Tomé (with Principe and Annobon). London: Brit. Mus. (Nat. Hist.).
Fairchild, D. (1931). A baobab tree in Florida. Nat. Hist. Mag. 10: 245–249.
Fanshawe, D. B. (1969). Forests Dept., Division of Forest Research, Kitwe, Zambia, pers. comm.
Fenner, M. (1980). Some measurements on the water relations of baobab trees. Biotropica 12: 205–209.
Ficalho, C. de (1884). Plantas uteis de Africa Portugueza. Lisbon.
French, M. H. (1944). Composition and nutritive value of pulp and seeds in the fruit of the baobab. E. Afr. Agric. J. 9: 144–145.
Fuchs, H. P. (1967). Pollen morphology of the family Bombacaceae. Res. Palaeobot. Palynol. 3: 119–132.
Galpin, E. A. (1968–69). Mosdene, Box 28, Naboomspruit, S. Africa, pers. comm.
Gardner, C. A. (1944). The vegetation of Western Australia. Journ. Roy. Soc. W. Austr. 28: xi–lxxxviii.
George, A. S. & Erdtman, G. (1969). A revision of the genus *Diplopeltis* Endl. (Sapindaceae). Grana Palynol. 9: 92–109.
Gerber, C. (1895). Contribution à l'histoire botanique, thérapeutique et chimique de genrre *Adansonia* (Baobab). Ann. Inst. Colon. Marseille 2, 2: 1–78.
Gillett, J. B. (1979). East African Herbarium, Ainsworth Hill, P.O. Box 45166, Nairobi, Kenya, pers. comm.
Gillman, C. (1949). A vegetation-types map of Tanganyika Territory. Geogr. Rev. 39: 7–37.
Greene, R. A. (1932). Composition of the pulp and seeds of *Adansonia digitata*. Bot. Gaz. 94: 215–220.
Grisard, J. (1891). Le baobab. Rev. Sci. Nat. Appl. 38(1): 76–78.
Guigonis, G. (1968). Directeur des Eaux-Forêts et Chasses, B.P. 830, Bangui, Central African Republic, pers. comm.
Guy, G. L. (1970). *Adansonia digitata* and its rate of growth in relation to rainfall in South Central Africa. Proc. & Trans. Rhod. Sci. Assoc. 54, 2: 68–84.
—— (1971). The baobabs: *Adansonia* spp. (Bombacaceae). Journ. Bot. Soc. S. Afr. 57: 30–37.
Harreveld-Lako, C. H. van (1926). *Adansonia digitata* L., de baobab of apenbroodboom. De Tropische Natur 15, 10: 157–162.
Harris, B. J. & Baker, H. G. (1959). Pollination of flowers by bats. Nigerian Field 24: 151–159.
Heel, W. A. van (1966). Morphology of the androecium in Malvales. Blumea 13: 177–394.
—— (1974). On dichotomy, with special reference to the funicles of the ovules of *Adansonia*. Proc. K. Ned. Akad. Wet. C., 77: 321–337.
Hiern, F. (1896). Catalogue African plants collected by Dr Friedrich Welwitsch in 1853–61. London: Brit. Mus. (Nat. Hist.).

Hobley, C. W. (1922). On baobabs and ruins. Journ. E. Afr. & Uganda Nat. Hist. Soc. No. 17: 75–77.

Hollis, R. (1963). Reflections on baobabs. Nigeria Field 28: 134–158.

Howard, W. J. (1976). Land resources of central Nigerian Forestry. Surbiton Land Resources Division, Land Resource Report 9.

Hughes, J. E. (1933). Eighteen years on Lake Bangweulu. London.

Hunting Technical Services. (1964). Report on the survey of geology, geomorphology and soils, vegetation and present land-use. Land and water resources survey in Kordofan Province of the Republic of the Sudan. Doxiades Assoc. Athens DOX–SUD–A26.

Ibn Batuta (1829). The Travels of Ibn Batuta. (trans. S. Lee). London.

Inamdar, J. A. & Chohan, A. J. (1969). Epidermal structure and stomatal development in some Malvaceae and Bombacaceae. Ann. Bot. 33: 865–878.

Irvine, F. R. (1930). Plants of the Gold Coast. London, Oxford Univ. Press.

Jaeger, P. (1945). Épanouissement et pollinisation de la fleur du Baobab. C. R. Acad., Sci. Paris 220, 11: 369–71; Fores Abstr. 8: 209 (1946).

—— (1950). La vie nocturne de la fleur de Baobab. La Nature No. 3177: 28–29.

—— (1954). Les aspects actuels du problème de la chéiroptèrogamie. Bull. I.F.A.N. A, 16: 796–821.

—— (1961). The Wonderful Life of Flowers (trans. J. P. M. Brenan). London.

Jahn, S. A. A. (1979). African plants used for the improvement of drinking water. Curare 2, 3: 183–199.

Johnson, M. A. (1961). On the shoot apex in the Bombacaceae. Amer. Journ. Bot. 48: 534.

Keay, R. W. J. (1949). An outline of Sudan zone vegetation in Nigeria. J. Ecol. 37: 335–364.

—— (1953). An outline of Nigerian Vegetation. Lagos: Govt. Printers.

——, Onochie, C. F. A. & Stanfield, D. P. (1960). Nigerian Trees, vol. 1. Lagos: Govt. Printers.

Kenton, R. H. & Legg, J. T. (1971). Varietal resistance of cacao to swollen shoot disease in West Africa. F.A.O. Plant Protection Bull. 19: 1–11.

Killick, D. J. B. (1969). Botanical Research Institute, P.O. Box 994, Pretoria, South Africa, pers. comm.

Kingdom, J. (1971). East African Mammals. An atlas of evolution in Africa, vol. 1. London.

Koch, D. (1972). Fruit-bats and bat-flowers. E. Afr. Nat. Hist. Soc. Bull. 1972: 123–126.

Koeleman, A. (1972). Transvaal's big trees. Flora & Fauna 23: 10–15.

Kranz, J., Schmutter, H. & Koech, W. (Eds.) (1977). Diseases, Pests and Weeds in Tropical Crops. Berlin.

Lafont, F. (1942). La croissance du Baobab. Notes Africaines No. 13: 8.

Lamprey, H. F. (1963). Ecological separation of the large mammal species in the Tarangire Game Reserve, Tanganyika. E. Afr. Wildlife Journ. 1: 12–92.

Lebrun, J.-P. & Stork, A. L. (1977). Index 1935–1976 des cartes du répartition des plantes vasculaires d'Afrique. Maisons Alfort.

Legg, J. T. & Bonney, J. K. (1967). The host range and vector species of virus from *Cola chamydantha* K. Schum., *Adansonia digitata* L. and *Theobroma*

cacao L. Ann. Appl. Biol. 60: 399–403.
Leo, J. (1600). A geographical historie of Africa written in Arabicke and Italian . . . trans. J. Pory. London.
Leriche, A. (1954). Autour du mot baobab. Notes Africaines No. 63: 89.
Lind, E. M. & Morrison, M. E. S. (1974). East African Vegetation. London.
Livingstone, D. (1857). Missionary Travels and Researches in South Africa. London.
Loret, V. & Poisson, J. (1895). Les végétaux antiques du Musée Egyptien du Louvre. Rec. Trav. Rel. Philolog. et Archéol. Égypt. et Assyr. 17: 177–199.
Loustalot, A. J. & Pagan, C. (1949). Local 'fever' plants tested for presence of alkaloids. El Crisol, San Juan, Puerto Rico 3(5): 3–5 [For. Abstr. 12: 229 (1950)].
Lubbock, J. (1892). A contribution to our knowledge of seedlings. London.
Lucas, G. L. (1971). The baobab map project. Mitt. Bot. Staatssamml. München 10: 162–164.
Maheshwari, J. K. (1971). The baobab tree: disjunction, distribution and conservation. Biol. Conserv.: 57–60.
Mauny, R. (1951). L'origine du mot baobab. Notes Africaines No. 50: 57–58.
—— (1955). Baobabs cimetièrs à Griots. Notes Africaines No. 67: 72–76.
Metcalfe, C. R. & Chalk, L. (1950). Anatomy of the Dicotyledons, vol. 1. Oxford.
Miège, J. (1960). Nombres chromosomiques de plantes d'Afrique occidentales. Rev. Cytol. Biol. Vég. 21: 373–380.
—— (1974). Étude du genre *Adansonia* L. II, Caryologie et blastogenèse. Candollea 29: 457–475.
—— (1975). Contribution à l'étude du genre *Adansonia* L. III, Intérêt taxonomique de l'examen électrophorétique des proteins des graines. Boissiera 24: 343–352.
—— & Burdet, H. M. (1968). Étude du genre *Adansonia* L. I, Caryologie. Candollea 23: 59–66.
Mogg, A. D. D. (1950). The Baobab. Trees in S. Afr. 1(4): 12–14.
Moll, E. J. & White, F. (1978). The Indian Ocean Coastal Belt. In M. J. A. Werger (Ed.) Biogeography and Ecology of Southern Africa: 561–598.
Muir, J. (1937). The seed-drift of South Africa and some influences of ocean currents on strand vegetation. Mem. Bot. Surv. S. Afr. No. 16.
Nachtigal, G. (1971). Sahara and Sudan IV, Wadai and Darfur (trans. A. G. B. & H. J. Fisher). London.
—— (in press). Sahara and Sudan II, Bornu (trans. A. G. B. & H. J. Fisher). London.
Newton, L. (1974). Is the baobab tree succulent. Cacts. & Succ. Journ. Gt. Brit. 36: 57–58.
Nicol, B. M. (1957). Ascorbic acid context of Baobab fruit. Nature 180: 287.
Nicholas, R.-P. F.-J. (1955). Recherches sur la valeur sémantique du mot Baobab. Notes Africaines No. 67: 77–78.
Nour, A. A., Magboul, B. I. & Kheiri, N. H. (1980). Chemical composition of baobab fruit (*Adansonia digitata* L.). Trop. Sci. 22: 383–388.
Oliver, B. (1959). Nigeria's useful plants. Nigerian Field 24: 13–34.
Owen, J. (1970). The medico-social and cultural significance of *Adansonia digitata* (Baobab) in African communities. African Notes 6, 1: 26–36.
—— (1974). A contribution to the ecology of the African baobab. Savanna 3: 1–12.

Palgrave, K. C. (1977). Trees of Southern Africa. Cape Town.
Palmer, E. & Pitman, N. (1972). Trees of Southern Africa, vol. 2. Cape Town.
Pardy, A. (1953). Notes on indigenous trees and shrubs of S. Rhodesia. *Adansonia digitata* L. (Bombacaceae). Rhod. Agric. Journ. 50: 5–6.
Pelly. R. G. (1913). The composition of the fruit and seeds of *Adansonia digitata*. J. Soc. Chem. Ind. 32: 778–779.
Perrier de la Bâthie, H. (1952). Sur les utilités de l'*Adansonia grandidieri* et les possibilités de culture. Rev. Int. Bot. Appl. Agric. Trop. 32: 286–288.
—— & Hochreutiner, B. P. G. (1955). Bombacacées. Flore de Madagascar, 130e famille, Bombacacées. Paris.
Pijl, L. van der (1936). Fledermäuse und Blumen. Flora 131: 1–40.
Popov, G. & Zeller, W. (1963). Ecological survey. Report on the 1962 survey in the Arabian Peninsula. F.A.O., Rome.
Porsch, O. (1935). Zur Blütenbiologie des Affenbrotbaumes. Oest. Bot. Zeit. 84: 219–224.
Posnette, A. F., Robertson, N. F. & Todd, J. McA. (1950). Virus diseases of cacao in West Africa, V. Alternative host plants. Ann. Appl. Biol. 37: 229–240.
Presting, D. (1962). Beiträge zur Pollenmorphologie madagassischer Pflanzenfamilien (mit einer taxonomischen Auswartung der Ergebnisse). Inaug. Diss. Christian – Albrechts Universität Kiel (typescript).
Radcliffe–Smith, A. (1979). Flora. In R. H. Daly (Ed.). Interim report and the results of the Oman Flora and Fauna Survey, Dhofar 1977, Sultanate of Oman: 41–48.
Rao, C. V. (1954). A contribution to the embryology of Bombacaceae. Proc. Ind. Acad. Sci. 39: 51–75.
Raven, P. H. & Axelrod, D. I. (1974). Angiosperm biogeography and post continental movements. Ann. Miss. Bot. Gard. 61: 539–673.
Ray, J. (1688). Historia plantarum. London.
Ridley, H. N. (1930). The dispersal of plants throughout the world. Ashford.
Riley, H. P. (1960). Chromosomes of some plants from the Kruger National Park. Journ. S. Afr. Bot. 26: 37–44.
Roberts, H. (1961). *Analeptes trifasciata* F., a longhorn beetle that attacks members of the Bombacaceae in Ghana. Extr. from Rep. W. Afr. Timb. Borer Res. Unit 1960; For. Abstr. 23: 102 (1962).
Robyns, A. (1963). Bombacaceae. Flore du Congo, du Rwanda et du Burundi 10: 191–204.
—— (1970). Un botaniste à Ceylon. Not. Belg. 51: 169–202.
—— (1972). Bombacaées et philatélie. Not. Belg. 53: 339–362.
Rock, T. D. (1861). Monkey bread nuts or fruit of the baobab. Technol. 1: 346–350.
Saccardo, P. A. (1898). Sylloge Fungorum, vol. 13.
Sahni, K. C. (1968). Important trees of the northern Sudan. Khartoum.
Scaliger, J. C. (1557). Exotericarum exercitationum liber quintus decimus de subtilitate ad Hieronymum Cardanum. Paris.
Schimper, A. F. W. (1903). Plant geography upon a physiological basis. Trans. W. R. Fisher, Oxford.
Schoenwald, H. R. (1969). [Clearing trial in savanna woodlands in the Blue Nile]. Forstarchiv. 40, 2: 21–25.
Schweinfurth, G. (1868). Pflanzengeographische Skizze des gesammten

Nil-Gebiets und der Uferländer des Rothen Meeres. Petermann, Mittheil 1868: 155–169, map 9.
—— (1912). Arabische Pflanzennamen aus Aegypten, Algerien und Jemen. Berlin.
Seabra, L., de (1948). Estudos de technologia. Contribuições paro o estudo technológico das espécies da flora colonial. Anais Junta de Invest. Colon., Lisboa 3(6): pp. 107.
Sinha, B. C. (1979). Tree Worship in Ancient India. New Delhi.
Société Horticole d'Alexandre (1901). Liste des plantes cultivées en Egypte. Alexandria.
Srivastava, G. S. (1959). Schizocotyly and polycotyly in *Adansonia digitata* Linn. Sci. & Cult. 25, 3: 218–219. For. Abstr. 21: 194 (1960).
Stahel, J. (1972). Eigenartiges Regenerationsvermögen des Baobab. Mitt. Deutsch. Dendrol. Ges. 64: 129.
Start, A. N. (1972). Pollination of the baobab (*Adansonia digitata* L.) by the fruit bat *Rousettus aegyptiacus* E. Geoffroy. E. Afr. Wildlife Journ. 10: 71–72.
Story, R. (1958). Some plants used by the Bushman in obtaining food and water. Mem. Bot. Surv. S. Afr. No. 30.
—— (1964). Plant lore of the bushmen. In D. H. S. Davies (Ed.). Ecological studies in southern Africa: 87–99. The Hague.
Strover, Mrs H. M. (1969–70). P.O. Box 17, Triangle, Zimbabwe, pers. comm.
Swart, E. R. (1963). Age of the baobab tree. Nature 198: 708–709.
Sweeney, C. (1969). Jebels by Moonlight. London.
Tarr, S. A. J. (1955). The fungi and plant diseases of the Sudan. C. M. I., Kew.
Thompson, J. V. (1816). A catalogue of the exotic plants cultivated in the Mauritius.
Thorold, C. A. (1975). Diseases of Cocoa. Oxford.
Tinsley, T. W. (1955). The host range of *Cacoa* viruses. Ann. Rep. W. Afr. Cocoa Res. Inst. 1954/55: 30–33.
Unwin, A. H. (1920). West African Forests and Forestry. London.
Vaid, K. M. (1964). Concluding chapter of a 'Kalpa-Vriksha'. Ind. For. 90: 1963–1964.
—— (1978a). Where is the mythical 'wishing tree'. Science Today, April 1978: 35–44.
—— (1978b). Temple and the tree. Wildlife Newsletter of Indian Forest College, Oct. 1978: 58–59.
—— & Ravinder Vaid (1978). Currency paper from *Adansonia*. Ind. Journ. For. 1: 53–55.
Varmah, J. C. & Vaid, K. M. (1978). Baobab, the historic African tree at Allahabad. Ind. For. 104: 461–464.
Vanguelin, L. N. (1822). Analyse des fruits du baobab, *Adansonia*. Mém. Mus. Hist. Nat. 8: 1–11.
Villiers, J.-F. (1975). Bombacaceae. Flore du Cameroun 19: 71–98.
Walker, A. (1953). La baobab au Gabon. Rev. Int. Bot. Appl. et Agric. Trop. 1953: 174–175.
Wallace, G. B. & Wallace, M. M. (1944). Supplement to the revised list of plant diseases in Tanganyika Territory. E. Afr. Agric. Journ. 10: 47–49.
Walter, H. (1964). Die vegetation der Erde in Ökologischer Betrachtung, 1. Die tropischen und subtropischen Zonen. Ed. 2. Jena.

Watt, J. M. & Breyer-Brandwijk, M. G. (1962). The medicinal and poisonous plants of southern and eastern Africa, Ed. 2. Edinburgh E. & S. Livingstone.
Watt, G. (1889). A dictionary of the economic plants of India. Calcutta.
Werger, M. J. A. (Ed.) (1978). Biogeography and Ecology of Southern Africa, 2 vols. The Hague.
—— & Coetze (1978). The Sudano–Zambesian Region. In M. J. A. Werger (Ed.). Biogeography and Ecology of Southern Africa: 301–462.
White, F. & Werger, M. J. A. (1978). The Guineo–Congolian transition to southern Africa. In M. J. A. Werger (Ed.). Biogeography and Ecology of Southern Africa: 599–620.
Wickens, G. E. (1977). The Flora of Jebel Marra (Sudan Republic) and its Geographical Affinities. Kew Bull. Add. Ser. V.
—— (1979). The use of the baobab (*Adansonia digitata* L.) in Africa. Proc. IX Plenary Meeting of A.E.T.F.A.T., Las Palmas de Gran Canaria, 18–23 March 1978: 27–34.
Wild, H. (1961). Bombacaceae. Flora Zambesiaca 1, 2: 511–517. London.
Williamson, J. (1955). Useful plants of Nyasaland. Zambia.
Willis, J. C. (1973). A dictionary of the flowering plants and ferns. Ed. 8, revised by H. K. Airy Shaw. Cambridge.
Woodruff, B. C. (1969). Baobab—Africa's tree of legend, myth and mystery. Sunday News Mag. [Tanzania], June 22nd: 7–8.
Wright, I. M. & Kerfoot, O. (1966). The African baobab—object of awe. Nat. Hist. N. York 75, 5: 50–53.

Printed in Great Britain by BPCC Wheatons Ltd, Exeter